Listening Lines:

Memory Currents

Published by New Dreaming Publications, August 2026

This book contains original writing and wisdom inspired by sacred traditions, water-based practices, ancestral teachings, and the lived experiences of the author. Any resemblance to existing works, traditions, or customs is offered with deep respect and acknowledgment to all, and especially for wisdom keepers past and present.

Cover design, illustrations, layout, and designs by Jenesis with collaborative support from contract graphic designers and artists, under her creative direction.

For rights, permissions, or to request community-based use of this material, please contact:

New Dreaming Publications
www.NewDreamingPublications.com

Direct correspondence to:
PO Box 544 Williamstown
Victoria, Australia 3016
T: +61 1300 722 599
E: info@NewDreamingPublications.com

ISBN: 978-0-9925428-9-4

Printed globally through Amazon.com

DISCLAIMER:

This book is edited commentary, interpretation and opinion. Much of the content is based upon scientific papers, personal experience and anecdotal evidence. It is meant to promote thoughtful consideration of ideas, spur philosophical debate, and inspire further independent research. It is not advice of any kind; don't use it to make decisions. It's not legal advice. It may contain errors and omissions, despite the Author's best efforts. New Dreaming Publications and Author, editors, copy editors, transcriptionists, designers and printers assume no responsibility for errors or omissions. Things can change quickly in the world; use this book as one reference, not your only reference.

References to any individuals, companies, products, and services are included for illustration purposes only and should not be considered endorsements.

Listening Lines:

Memory Currents

BELONGING

A Devotion to Port Phillip Bay

"This book will not tell you what the bay is.

It will change how you meet it."

AUTHOR'S NOTE

On Listening, Place, and What Emerges

This work began without a plan to explain anything.

It began with attention to whales. And with returning to the shoreline, of the waters of Port Phillip Bay.

To a practice of sitting, watching, and listening without needing an immediate outcome.

Over time, something began to reveal itself. Not as answers, but as patterns. Not as conclusions, but as a deepening awareness that place is not fixed - that what we see at the surface is only one layer of something far more continuous.

Listening, in this sense, is not passive. It is a form of attention that allows what is already present to become visible.

Listening Lines are not drawn on maps. They are formed through return - through attention, through the body, through a relationship with place that develops over time. They do not appear all at once. They emerge slowly.

I respect the science, reference it as carefully as I can, and then interpret it as responsibly as I can. This work does not seek to replace scientific understanding, but to sit alongside it - observing, connecting, listening and translating what is known into a way of seeing grounded in place, experience, and attention.

I am a bridge-builder between knowledge and meaning.

This book does not attempt to define what cannot yet be fully known.

It does not claim ownership of what belongs to the land, the water, or the stories that exist beyond individual experience.

Instead, it offers a way of noticing.

A way of being present at the edge of something - and allowing that edge to remain open.

What follows is not a conclusion.
It is an invitation.
To sit.
To observe.
To listen.
And to allow what is already there to be seen, perhaps for the first time.

Enjoy.

Jen xo

Table of Contents

A Quiet Arrival

You arrive at the edge of the bay.
Not announced.
Not marked.
Just the meeting of water and sky, held in a line that shifts as you watch it.

The air carries salt.
Soft, familiar, moving across the open stretch of water and into the body without asking.
There is space here.
Not empty.
But wide.

Behind you, things of the city continue. Movement. Sound.
The quiet persistence of a place lived in.
In front of you, the water holds a different pace.

You stand, or sit, or pause along the shoreline. Rocks. Sand. Timber worn by time and tide. Nothing needing attention. And yet, something is happening.

The movement you brought with you loosens. Not all at once. Just enough.

Light shifts across the surface of the bay. Wind moves through it, then settles. The body, often unnoticed, begins to register what is already here.

You do not need to understand it. You do not need to name it.

Only this is asked - Stay.

Long enough for the edge to stop feeling like a boundary.
Long enough for something quieter to come forward.

This is where it begins.

Introduction - Where Memory Moves

There is a way of being at the water's edge that does not begin with looking. It begins with staying.

Not for long at first. Not with intention. Just long enough for something to shift.

The surface settles. The need to interpret softens.

The body, which has been moving quickly through the world, begins to register something else.

Sound carries differently here. Light behaves differently. Time does not move in the same way.

It is not that the bay changes. It is that attention does.

This work did not begin as a theory. It began as a return. To the same stretch of shoreline. Again, and again.

At different times of day. In different conditions. With no fixed outcome.

To sit.
To watch.
To listen.
And, at times, to enter the water.

Over time, a pattern became visible.

Not in the way we are used to seeing patterns - through data, measurement, or recorded sequence - but through repetition and recognition.

Something continued.
Across tides.
Across weather.
Across days that felt entirely different on the surface.

There was a consistency beneath variation. A movement that did not begin or end with any single moment of observation.

This book reveals how memory moves - through water, through Earth, through life.

Water does not store memory in the way we often imagine. But it carries pattern, movement, and influence forward through time.

Memory is not held. It is expressed. It moves through pattern, through repetition, through response across systems.

It can be observed in:

- ocean currents
- whale migration
- geological record
- human sensory recognition.

What we call memory is not something kept. It is something that continues.

This is what I have come to understand as **Memory Currents**.

To notice this requires a different form of attention. Not one that extracts meaning nor that seeks conclusion. But one that remains. One that allows perception to reorganise over time. This is not passive. It is a discipline. A practice.

Whales are not incidental to this work. They do not simply appear as moments of wonder or interruption. They live within the very systems this book is observing.

They move through vast oceanic pathways with a consistency that extends across generations. They respond to temperature, pressure, sound, and subtle shifts within the water. Their migrations follow patterns that are carried, continued, repeated. This is memory, expressed through life. Not stored. Not retrieved. But lived.

They do not remember in the way we think of memory. They continue in it.

In this way, whales are not separate from Memory Currents. They are among its clearest expressions.

They embody a form of listening that is not intellectual, but cellular. A listening that occurs through the body, through orientation, through response to the conditions of the world as they are.

This is **embodied listening**. The practice that emerged alongside this understanding is what I have come to call **Listening Lines**. They are not drawn on maps.

They are not visible in the way we are used to seeing.

They are formed through return. Through attention.

Through the body.

Through a relationship grown slowly over time.

They are not followed. They are felt.

Along the shoreline of Port Phillip Bay, and more specifically the waters off Williamstown, this practice took shape.

Not as something formal. Not as something taught. But as something that became unavoidable once it was noticed.

This is not a book about *proving*. It is a book about *noticing* what is already occurring. About remaining long enough to perceive it. And recognising that what we call memory is not behind us. It is moving. And we are within it.

At the Edge of the Bay

There is a place along the edge of the bay where nothing in particular is happening.

No event. No marker. No reason to stop, beyond the simple fact that you do.

Along the shoreline of Williamstown, the bay opens wide. Not enclosed. Not held tightly.

It stretches outward, carrying light, wind, and movement across its surface in ways that are never exactly the same, and yet never unfamiliar.

At first, it appears as it always has. Water moving. Wind passing. Light shifting. Nothing asking for attention.

And yet, something begins to register. Not through thought. Through the body.

The pace you arrive with does not belong here.
It takes a moment to notice that.
Longer still to allow it to settle.

There is a tendency to look for something.
A point of focus.
A reason to remain.
Something to confirm that this moment holds meaning.

But the bay does not offer itself in that way.
It does not present.

It does not perform.

It continues.

The surface moves with the wind, then without it.
The tide shifts in ways too gradual to follow in real time.
Cloud passes.
Light changes.
The horizon holds, then softens.

Nothing announces itself.
And yet, if you remain, something becomes unmistakable.

There is pattern here.
Not fixed.
Not repeating in exact form.
But consistent.
A continuity beneath variation.

It is not something you see all at once.
It gathers.
Through staying.

The body begins to recognise what the mind has not yet named.
A subtle orientation.
A familiarity without reference.

This is where the shift occurs.
Not in the water.
In perception.

What was previously background, begins to hold presence.
What was overlooked, begins to organise itself differently.

Not into answers. Into awareness.

You begin to notice what continues.
Across moments that seem unrelated.
Across changes that appear separate.

The same movement, carried differently.
The same response, expressed through variation.

This is not observation as we are used to it.
It is not about identifying, categorising, or concluding.

It is about remaining long enough for pattern to reveal itself.

The edge of the bay is not a boundary. It is a threshold.

A place where what appears still and what is constantly moving meet. A place where perception can shift, if given the time.

Nothing has been explained.
Nothing has been proven.

And yet, something is known.

Not through instruction.
Not through analysis.

Through recognition.

This is where Listening Lines begin.
Not as something seen.
But as something felt.

They are not laid out. They are not followed.

They emerge. Through return. Through attention. Through the body learning, slowly, to remain.

There is no moment where this begins clearly. No point at which you can say - *Now I understand.*

Only this - You come back. You stay. You notice more than you did before.

And over time, what once felt like a place becomes something else.

Not separate from you. Not outside of you. But something you are already within.

The Practice of Staying

It is easy to arrive. It is far less common to remain.

At the edge of the bay, the first moments require very little. You stand. You look. You take in what is there. This is familiar.

It fits within the way we have learned to move through places - to notice quickly, to register, and then to continue.

Staying interrupts that pattern.

There is no clear instruction for how to do it. No measure of success. No visible shift that confirms you are "doing it right."

And yet, it is the only way anything deeper becomes available.

To stay is not to hold yourself in place through effort. It is not endurance. It is not waiting for something to happen.

It is a softening of the impulse to leave.

At first, the body resists. It looks for cues.

How long should this take? What am I meant to notice? When will something change?

These questions arise quickly. They belong to a pace that does not originate here.

The bay does not respond to them.

Nothing accelerates to meet your expectation. Nothing presents itself more clearly because you are looking harder.

It continues as it is.

This is where the practice begins.

Not in what you observe, but in how you remain.

The body begins to shift in small, almost imperceptible ways. Breath settles. Vision softens. Attention widens, then narrows, then widens again.

What felt like stillness reveals movement. What felt like emptiness reveals detail.

Not because anything has changed. Because you have.

Time behaves differently here.

Not slower. Not faster. Less defined.

Minutes are not marked in the same way.

They stretch. They fold. They pass without the usual signals that tell you something has been completed.

This can feel unfamiliar. Even uncomfortable.

There is a point, often early, where the impulse to leave returns with force. It can feel practical. Reasonable.

You have things to do. Places to be. No clear reason to remain.

This is the moment most people leave.

Not because nothing is happening. But because nothing is happening in the way they expect.

If you stay beyond this point, something shifts.

Not dramatically. Not all at once.

The need for outcome loosens. The sense of time reorganises. The body stops asking for direction.

Attention begins to move differently.

Not searching. Not scanning.

Receiving.

You begin to notice what continues. Not as isolated moments, but as something carried.

A pattern that does not repeat exactly, and yet is unmistakably familiar.

The water does not present this to you. It does not highlight it.

It is simply there.

The longer you remain, the less you feel separate from what you are observing.

The boundary between watcher and movement softens.

You are not outside of it, looking in. You are within the same conditions. Responding, even if subtly, to what is occurring.

Wind touches the surface of the water. It touches you.
Light shifts across the bay. It shifts across your field of vision.

The same conditions are moving through both. This is where the practice deepens.

Staying is no longer something you are doing. It becomes something you are part of.

There is no need to hold it. No need to understand it.

Only to continue.

Return makes this possible.

One visit does not reveal much.
Two visits begin to feel familiar.

Over time, the body recognises the place before the mind does.

The same stretch of shoreline. The same openness of the bay.
The same subtle shift as you arrive.

What once felt undefined begins to organise itself through repetition.

Not into knowledge. Into recognition.

This is how Listening Lines begin to form.

Not through a single moment of clarity. But through accumulated presence.

Through staying long enough, often enough, that what was once unnoticed becomes impossible to overlook.

There is no endpoint to this practice. No moment of completion. Only a deepening.

You return. You remain. You notice more than you did before.

And gradually, without announcement, something changes. Not in the bay. In your capacity to perceive it.

Listening without Taking

There is a way of paying attention that reaches outward. It gathers. It interprets. It makes something useful from what is observed.

This way of looking is familiar. It has been practised over time, shaped by systems that reward understanding, explanation, and outcome.

It is not wrong. But it does not belong here.

At the edge of the bay, this way of listening begins to feel out of place. It shows itself quickly. In the impulse to name what you are seeing. To understand the movement of the water. To assign meaning to what appears.

The mind reaches.

What is that pattern? What does that shift mean? Is this important?

These questions move ahead of perception. They arrive before anything has had time to settle. The body tightens around them. Attention narrows. The field of what can be noticed becomes smaller, more defined.

This is the moment where listening turns into taking.

Something is extracted. Pulled into thought. Held, briefly, as something known.

And then it is gone.

The bay does not resist this. It continues. But something is lost in the process. Not information. Presence. To listen without taking is to allow what is perceived to remain as it is.

Not captured. Not translated immediately into meaning.

It requires a different orientation. Instead of reaching outward, attention settles. Instead of narrowing, it widens. Instead of asking what something means, it allows it to exist without explanation.

This is not passive. It is not disengaged. It is a form of restraint. The restraint not to interfere with what is being perceived.

At first, this feels unnatural. The impulse to name, to understand, to hold onto something concrete returns again and again. It can feel as though nothing is happening if nothing is being concluded.

But something is happening. Perception is reorganising. Without the constant interruption of interpretation, the body begins to register more.

Subtle shifts. Small continuities. Patterns that do not announce themselves.

What was previously unnoticed becomes available. Not because it was hidden. Because it was not being taken.

Listening without taking allows what continues to remain in motion. It does not fix it in place. It does not reduce it to something static. It recognises that what is being perceived is not an object. It is movement.

This is where the distinction becomes clear. Taking creates distance. It places what is observed outside of you. Something to be understood, analysed, and held.

Listening dissolves that distance. You are no longer separate from what is occurring. You are within the same conditions.

Wind moves across the water. It moves across you.

Light shifts across the surface. It shifts across your perception.

The same field holds both. Nothing has been taken. And yet, more is known.

Not as information.
As recognition.

This form of listening cannot be rushed.

It does not respond to effort. It responds to presence.

The longer it is practised, the less the impulse to take arises.

Not because it has been suppressed. Because it is no longer needed.

Understanding begins to emerge in a different way. Not constructed. Not assembled.

Felt.

This is where Listening Lines begin to clarify.

Not as something you follow. Not as something you define.

But as something you are already within, once the need to take has fallen away.

There is nothing to hold.

Only this – Remain. Listen. Let what is moving continue.

The Body as Instrument

There is a point where listening is no longer something you do. It is something that occurs.

Not through effort. Through the body.

At the edge of the bay, after staying, after loosening the impulse to take, something becomes more apparent.
You are not only observing.
You are registering.

This distinction is subtle at first.

Observation keeps a distance.
It looks outward.
It gathers what can be seen.

Registration happens differently.

It does not begin with sight.

It begins within the body.

A shift in breath.
A change in orientation.
A quiet adjustment in posture that you did not consciously choose.

The body is already responding.

Before thought forms. Before meaning is assigned.

This is not unusual. It is simply unnoticed.

The body is constantly receiving.

Temperature.
Movement.
Pressure.
Sound.

Not as separate inputs. As a field.

At the shoreline, this becomes easier to feel.

The wind does not only move across the surface of the bay. It moves across your skin.

The same current that shifts the water alters your balance, your stance, your breath.

Light changes.
Your eyes adjust.
Your body follows.

Nothing is happening in isolation.

You are not outside of the conditions you are observing.

You are within them.

This is where the body reveals itself as an instrument.

Not something separate from perception.
Not something that receives after the fact.

But something that is actively sensing, orienting, and responding in real time.

The body does not wait for instruction.

It recognises.

Before there is language. Before there is thought.

There is a form of knowing that occurs through sensation.

A familiarity that cannot be traced to a single moment.

A recognition that arrives without explanation.

This is often overlooked.

Because it does not present as clear information.

It does not say - *This is what is happening.*

It says nothing.

And yet, it guides.

A slight shift in where you stand.

A subtle pull to remain longer than you intended.

A sense that something has changed, even when you cannot point to what.

These are not random. They are responses.

At a cellular level, the body retains patterns of interaction with its environment. Not as stored stories.

But as readiness.

As the capacity to recognise and respond to conditions that have been encountered before.

Through processes studied in fields such as neuroscience and epigenetics, it is understood that experience shapes how the body functions over time.

Response becomes patterned. Recognition becomes faster. Orientation becomes more precise.

In this work, it is felt more simply.

The body remembers through response.

At the edge of the bay, this becomes more noticeable.

Not because the bay is doing something extraordinary.

Because you are allowing yourself to register what is already occurring.

The longer you remain, the more the body begins to take precedence over the mind.

Not replacing it.
Rebalancing it.

Thought slows. Sensation leads.

You begin to trust what is felt before it is explained.
Not blindly. Not without awareness.

But with a recognition that the body is not secondary.

It is primary.

This is where embodied listening deepens.

It is no longer about trying to perceive something subtle.

It is about allowing perception to occur through the systems that are already designed for it.

The body does not need to be taught how to do this.

It needs to be allowed.

The conditions are already present.

Water moving.
Wind shifting.
Light changing.

And within those same conditions:

Your breath adjusting.
Your stance shifting.
Your awareness widening.

There is no separation between these.
They are part of the same field.

This is what makes the body an instrument.

Not because it is used.

But because it is already in relation.

Already responding.
Already participating.

Nothing needs to be added.

Only noticed.

And over time, through return, through staying, through listening without taking, this becomes more familiar.

Not as knowledge.

As orientation.

You begin to recognise when you are in it.
When your perception has shifted.
When the body is leading.

There is no clear boundary.

No moment where it begins definitively.

Only this - You arrive. You stay. You listen.

And gradually, without announcement, the body takes its place as the primary way of knowing. Not separate from the bay. Part of the same movement.

How Memory Moves

This book reveals how memory moves - through water, through Earth, through life. It does not begin where we are used to looking for it.

Not in storage. Not in record. Not in something held and retrieved.

What is being observed here does not sit still.

It continues.

At the edge of the bay, this becomes perceptible. Not immediately and not in a single moment. But through return.

Through staying long enough that what first appears as change begins to reveal continuity.

The surface shifts constantly.
Wind alters direction.
Light refracts differently across the water.
Tide moves in and out.

Nothing appears the same.

And yet, something unmistakable persists.

Not as repetition in exact form.
But as pattern.

A way in which movement carries forward.

A way in which response continues.
This is where the idea of memory begins to shift.

Memory is often understood as something stored. Held in place. Accessible as something that has already occurred.

But what is present here does not behave like that.

Nothing is retrieved.

Nothing is brought back from a fixed point in the past.

Instead, what has occurred continues to influence what follows.

Conditions carry forward.
Movement carries forward.
Response carries forward.

Not as record. As continuity.

In ocean systems, this is visible in the persistence of currents.

Water does not simply move and stop.
It carries temperature, salinity, and motion across vast distances and timeframes.

What occurs in one place influences what unfolds elsewhere.

Not as stored information.

But as ongoing movement.

In living systems, this continuity takes other forms.

Migration pathways followed across generations.
Behaviour that persists without instruction.

Patterns that are not taught in the way we understand teaching. But carried. Continued. Repeated.

The same can be felt within the body.

Recognition that arrives before thought.
Orientation that occurs without deliberate reasoning.

A response that feels immediate, but is not without history.

Not stored as narrative.

Held as readiness.

Across these systems - oceanic, biological, sensory - something consistent is present.

Memory is not fixed.

It moves.

It is expressed through pattern.
Through repetition.
Through response.

This is what is being named here as **Memory Currents**.

Not as a defined object. Not as something measurable in a single way.

But as a way of describing how continuity is carried through living and physical systems.

At the bay, this is not something you prove.

It is something you begin to notice.

The longer you remain, the more what once appeared as separate moments begins to feel connected.

Not because you have constructed a link. Because you are perceiving what continues.

The water does not return to a previous state.

It moves forward.

And yet, within that movement, there is familiarity.

A pattern that does not repeat exactly.
But does not disappear.

This is not memory as something behind you.

It is memory as something you are already within.

The past is not held in place.
It is carried forward through the conditions that remain in motion.

This is why it cannot be seen all at once.

It must be felt across time.

Through return. Through attention. Through the body learning to recognise what continues, even as everything appears to change.

Nothing has been stored.

And yet, nothing is lost.

It is all moving.

And you are already part of it.

Pattern, Repetition, Continuation

What moves does not move randomly.

At first, this is difficult to see.

The surface of the bay appears in constant change.
No two moments match.
No pattern repeats in a way that can be easily named.

It can feel as though everything is shifting without order.

But this is only what is visible at a glance.

When you remain, something else begins to emerge.

Not a fixed pattern. Not something that can be mapped or predicted with precision.

But a consistency.

A way in which movement carries itself forward.

The wind does not move the same way twice.
And yet, there is a familiarity in how it arrives, how it passes, how it alters the surface of the water.

The tide does not present itself as a repeated event.
And yet, it continues.

Not identical.
But continuous.

This distinction matters.

Repetition is often understood as something that returns in the same form.

A cycle. A loop.

But what is present here is different.

Nothing returns exactly as it was.

And yet, nothing begins entirely new.

This is pattern without exact repetition.

A continuity that allows variation.

A movement that adapts while remaining recognisable.

This is how memory moves.

Not by recreating the past.

But by carrying forward what has already shaped the system.

Each movement is influenced by what came before.

Not as a copy.
As an extension.

This can be seen across the bay.
The way the surface responds to wind is shaped by the conditions already present.

The direction, the temperature, the subtle movement beneath what is visible.

Nothing acts alone.

Each moment carries the influence of what has already occurred.

And yet, it does not repeat it.

It continues it.

This is the difference between repetition and continuation.

Repetition suggests return.

Continuation suggests movement.

At the shoreline, this becomes something you begin to recognise.

Not through counting or tracking.

Through familiarity.

A sense that what is occurring now belongs to something that has been unfolding all along.

This recognition does not arrive as certainty.

It arrives as orientation.

You begin to feel when something fits within the ongoing movement. And when it does not.

Not as judgment.

As resonance.

This is subtle.

It does not announce itself clearly.

But over time, it becomes unmistakable.

The body begins to align with what continues.

To recognise the difference between what is part of the pattern and what is momentary surface.

This is not about separating the two.

Both are present and both are real.

But one passes quickly. The other carries forward.

This is where attention deepens.

Not toward what stands out.

Toward what persists.

The surface will always change.

It will always offer variation, movement, distraction.

But beneath that, there is something that does not disappear.

Not fixed.

But continuous.

This is where Listening Lines begin to hold.

Not as lines that are drawn.

But as pathways of recognition.

Formed through repeated presence.

Through returning to the same place, the same conditions, the same edge between what appears and what continues.

Each time, something is added.

Not as information.

As familiarity.

As the body learning to recognise what does not need to be explained.

This is how the lines form.

Not through a single moment.

Through accumulation.

Through repetition that is never exact, but always connected.

Through continuation that carries forward what has already been set in motion.

There is no final pattern to arrive at.

No complete understanding to reach.

Only this - You begin to recognise what continues. And once recognised, it cannot be unseen.

Even as everything else continues to change.

Systems That Remember

No movement exists on its own.

At the edge of the bay, what appears as a single field of water is not singular. Wind moves across the surface.

Tide shifts beneath it. Temperature layers through depth.

Currents carry influence from beyond what can be seen. Each of these moves according to its own conditions, and yet none of them are separate. They interact.

They respond. They shape one another continuously.

What is visible is only the surface of that interaction. What is occurring extends far beyond it.

This is where memory becomes more clearly understood as something distributed. Not located in a single place. Not held within a single system. Carried across relationship.

Ocean systems do not operate in isolation. What occurs in one region influences another. Temperature shifts travel. Currents extend. Conditions persist beyond their point of origin. Not as stored events, but as ongoing influence.

The bay is part of this. Not closed. Not separate. Connected.

What enters it has already been shaped by what came before. What leaves it carries that influence forward.

Nothing begins here. Nothing ends here. It is part of a wider field.

This is true beyond water.

Living systems operate in the same way. No organism exists in isolation from its environment. Response is shaped by interaction. Behaviour emerges from relationship. Patterns are carried not by a single body, but across generations, across conditions, across time.

This is how migration persists. Not as a fixed instruction stored in one place, but as a pattern maintained across a system. Through repetition. Through response. Through continuation.

The same is true within the human body.

No single cell holds the entirety of experience, and yet the body responds as if it recognises. Not as stored narrative, but as patterned readiness across systems. Nervous system. Sensory perception. Cellular response. Each contributing. Each carrying forward influence. No single point of storage. Only participation.

This is what it means for a system to remember. It does not recall. It continues.

At the bay, this becomes something you begin to feel. Not as a concept, but as a condition. You are not observing separate elements. You are within an interacting field.

Wind does not move independently of water. Water does not move independently of tide. Tide does not move independently of larger ocean systems. And you are not separate from any of it.

The same conditions that move the water move through you. Temperature. Pressure. Sound. All received. All responded to.

You are part of the system.

Not observing it from the outside. Participating within it. This is where the idea of separation begins to loosen. Not as a belief, but as a direct experience. The boundary between observer and environment is no longer fixed. You begin to feel the continuity between what is happening around you and what is happening within you.
The same field. Different expressions.

This is where Listening Lines deepen. They are not lines between points. They are pathways within a system. Felt through relationship. Recognised through participation. Not followed. Lived.

Nothing in this system is static. And yet nothing is lost. Everything continues, carried across interaction. Not held in place, but held in motion.

This is how systems remember.

The Bay as Living Record

The bay does not hold memory in the way we are used to understanding it.

It does not store. It does not preserve in fixed form. And yet, something continues.

Patterns return.

Movements repeat.

Life follows pathways that extend beyond a single moment.

This is not immediately visible. From the shoreline, the bay can appear still. Even uneventful.

But with repeated attention, something begins to emerge. Not as information. But as continuity.

The bay is not simply a place where events occur.

It is a system in which movement, presence, and change are expressed over time.

Currents move through it. Tides shape it. Species enter and leave. Conditions shift.

And within that, patterns persist. Not perfectly. Not identically. But recognisably.

To call it a record is not to suggest that it holds memory in a fixed or retrievable way.

There is no archive beneath the surface. No stored history waiting to be accessed.

What exists instead is continuation. An ongoing expression of movement and relationship.

A record that is not written, but lived.

This becomes apparent only through return.

Not through a single visit. Not through observation alone. But through repeated presence.

Through watching over time. Through entering the water. Through noticing what remains, even as everything shifts.

In this way, the bay does not reveal itself all at once. It becomes known gradually.

Through accumulation. Through familiarity. Through the body learning to recognise what continues.

This does not require interpretation. It does not require belief.

Only attention.

The idea of a living record does not sit apart from what is observed.

It arises from it.

From the movement of water. From the behaviour of species. From the repetition of patterns across time.

Along this coastline, there is also evidence of presence that extends far beyond the present moment.

Fossil records indicate that whale species once inhabited these waters in distant geological time.

Today, other whale species move along the same broader coastline.

These are not presented as direct continuities of behaviour. But they do indicate something else.

That this place has long been part of larger systems of movement. What enters the bay now does not arrive in isolation. It arrives within a context that extends beyond what is immediately visible.

To describe the bay as a living record is not to explain it. It is to recognise that what is observed here is part of something ongoing.

Migration as Memory in Motion

Movement does not begin at the point it is seen.

When a whale passes through the bay, it is already within a much larger trajectory. What is visible from the shoreline is only a fragment of a movement that extends across distance, across seasons, across generations.

The path is not created in that moment. It is continued.

Migration, as it is commonly understood, suggests a journey from one place to another. A departure. An arrival. A route between two defined points.

But what becomes apparent over time is that migration is not simply movement between locations. It is continuity expressed through movement.

The whales that move along this coastline are not navigating a new path each season. They are continuing pathways that have been carried forward, maintained through repetition that is never exact, but always connected.

These pathways are not marked in the water.

They are not visible to the eye.

And yet, they are followed.

Not as instruction.

As orientation.

The body knows where to move. Not through conscious decision, but through attunement to conditions that extend beyond immediate perception.

Temperature gradients, ocean currents, geomagnetic fields, acoustic pathways carried through water.

Each of these contributes to a field of guidance that is not fixed in one place, but distributed across the environment.

The whale does not hold a map. It moves within a system that is already informing its direction.

This is where migration becomes inseparable from memory.

Not as something stored and recalled, but as something carried through the interaction between body and environment.

The pathway exists because it is continued.

It is maintained not by being held in place, but by being moved through.

Each generation does not begin again. It enters into what is already in motion. This is what gives migration its continuity. Not repetition in the sense of duplication, but repetition as participation in an ongoing movement.

The route is not fixed. Conditions shift. Water temperature changes. Currents alter. Sound travels differently. And yet, the movement persists.

Not because it remains identical. Because it adapts while continuing.

This is memory in motion.

At the edge of the bay, this is difficult to see all at once.

What is visible is a brief passage.

A moment where something vast crosses into view.

But that moment is connected to a movement that extends far beyond what can be observed directly.

The whale is not arriving for the first time.

It is passing through.

Not only through space.

Through continuity.

What it carries is not something contained within its body alone.
It is part of a system that extends across oceanic scale.
A system where movement is informed by what has already been carried forward.

This is why migration cannot be reduced to instinct as something fixed.

It is responsive. Dynamic.

Shaped by conditions that are constantly changing, yet remain connected through underlying pattern.

The whale moves within this field. Responding to it. Continuing it.

At the shoreline, something begins to shift in how this is perceived. The idea of distance changes.

What seemed far becomes present. What seemed separate becomes connected. The bay is no longer an isolated location. It is part of a pathway.

Part of a movement that extends beyond what can be seen, but is not separate from what is here.

This is where Listening Lines deepen again. They are not limited to a single place. They extend through continuity.

Through pathways that are not visible, but can be felt through the way movement carries itself forward.

You begin to sense that what is occurring here is connected to what is occurring elsewhere.

Not abstractly.
Directly.

The same water systems. The same currents. The same conditions influencing movement across distance. The same continuity.

Migration reveals this. Not as explanation. As experience.

You begin to feel that what you are witnessing is not an isolated event, but a moment within an ongoing movement.

A movement that did not begin here. And does not end here. This is where the scale of Memory Currents becomes more fully realised. They are not contained within a single location.

They move across systems. Across distance. Across time. Carried through interaction, through response, through continuation.

The whale does not travel through empty space. It moves through a field that is already structured by what has come before.

A field that informs its movement even as it contributes to it.

This is the nature of continuity.
It is not static.
It is not fixed.
It is carried forward through movement itself.

At the edge of the bay, this can only be sensed in fragments. Moments where something vast intersects with what is immediately present.

But even in those moments, something becomes clear.

What you are witnessing is not arrival. It is continuation. And in recognising this, your own perception begins to shift.

You are no longer seeing a single movement. You are sensing a pathway.

Not laid out. Not defined. But carried.

Through water.
Through Earth.
Through life.

And you are already within it.

Embodied Listening

Listening does not begin with the ear.

It does not begin with sound.

It begins with the body.

At the edge of the bay, this becomes more apparent over time. Not immediately. Not as a concept. But as a shift in how perception occurs.

What you notice is no longer limited to what you can see or name. It begins to include what you can feel without explanation.

A change in air pressure before the wind arrives. A subtle adjustment in balance as the ground beneath you settles. A shift in breath that happens before you are aware of it.

These are not separate events. They are responses. The body is already in relation to what is occurring.

Long before thought forms, before language arranges experience into meaning, the body has registered something.

This is where listening changes. It is no longer something you direct outward. It is something that occurs through you.

The senses do not operate in isolation.

They move together.

Sight, sound, touch, orientation, internal sensation - each contributing to a field of perception that is continuous rather than divided.

At the shoreline, this field becomes easier to recognise.
The movement of water is not only seen.
It is felt through rhythm.

Through timing. Through the way the body begins to align with what continues.

This is not something learned through instruction. It is something that becomes available when the pace of perception changes.

When the need to identify, interpret, and conclude softens.

The body begins to lead. Not in opposition to the mind. In advance of it.

Recognition arrives before explanation. Orientation occurs before decision.

There is a knowing that does not require language to exist. This is often overlooked because it does not present as clear information.

It cannot be easily described. It does not say what something is. It indicates how something is.

This distinction matters.

To know what something is requires separation. To sense how something is requires participation.

Embodied listening does not separate you from what you are perceiving. It places you within it.

At the edge of the bay, this becomes increasingly clear. Wind moves across the water. It moves across you.

The same movement. Different expressions.

Light shifts across the surface. It shifts across your field of vision.

The same condition. Different forms.

You are not outside of what is occurring. You are within the same field. Responding continuously, whether you are aware of it or not.

Embodied listening is the act of becoming aware of that response.
Not controlling it.
Not analysing it.
Not trying to improve it.
Simply recognising that it is already happening.

The body does not need to be trained to do this. It needs to be allowed.

When attention settles, when the impulse to take and interpret loosens, the body's capacity to register becomes more apparent.

Subtle patterns begin to stand out. Not as objects. As continuity.

You begin to sense when something is aligned with what is already moving.

And when something is not.

This is not judgment.
It is resonance.

The body recognises what fits within the ongoing field.
This is the same principle that underlies migration.

The whale does not calculate its path in the way we might imagine.

It responds to a field of conditions.
It aligns with what continues.
It moves in relation to that.

Embodied listening in the human does not replicate this.
But it moves in the same direction.
Away from separation.
Toward participation.

You begin to trust what is felt before it is explained.
Not blindly.
Not without awareness.
But with a recognition that perception does not begin with thought.

It begins with contact.
With relationship.

With the body already in response to what is present.

At the bay, this is not something that needs to be forced. It emerges. Through staying. Through returning. Through allowing perception to reorganise over time.

There is no clear point where this begins.

No moment where you can say that you are now listening in this way. Only a gradual shift.

You notice more. You feel more. You respond differently.

Not because you have learned something new.
Because you have stopped interrupting what was already there.

Embodied listening is not an addition.
It is a removal.
Of distance.
Of interference.
Of the need to stand outside of what is occurring.

What remains is simple.
You are here.
The bay is here.

The same conditions move through both. And in that shared movement, something becomes clear.

You are not separate from what you are perceiving.
You are part of it. Already listening.

Whether you realised it or not.

When Whales Enter the Bay

There are times when the bay changes. Not in surface. Not in weather. Not in anything immediately visible.

But in presence.

The first sign is not always seen. It is felt. A shift that is difficult to locate precisely, but unmistakable once recognised. The field alters. Attention sharpens without effort. The body becomes more aware, not in tension, but in response.

Something has entered.

At times, it becomes visible. A movement breaking the surface. A slow rise. A breath. A return beneath.

A whale.

But even this is not the beginning.

By the time the whale is seen, it is already within the system. Already responding. Already participating in the same conditions that move the water, the wind, the light. Its presence is not separate from the bay. It is an extension of it.

The whales that move through these waters are most often baleen whales, travelling along ancient migratory pathways that extend far beyond the visible horizon.

Their movement is not confined to this place, but this place is part of their movement.

They do not arrive randomly.

They continue.

Across seasons. Across generations. Across distances that cannot be held in a single moment of observation.

Their pathways are not taught in the way we understand teaching. They are carried. Expressed through movement.

Maintained through continuation.

This is memory in motion.

Not stored. Not retrieved.

Lived.

The movement of the whale does not sit on top of the water. It moves within it, guided by conditions that are not immediately visible but deeply felt. Temperature gradients.

Pressure changes. Sound travelling across distance. Subtle variations in the water that do not register to the eye, but are unmistakable to a body attuned to them.

This is not instinct as something fixed. It is response within continuity.

When a whale enters the bay, it does not bring something separate into the system. It reveals what is already there.

The bay is not isolated. It is part of a larger field of movement that extends along the southern coastline and beyond. The whale makes this visible.

Not as information.

As presence.

What once felt contained begins to open. What seemed local reveals itself as connected.

The scale shifts.

The sense of distance changes.

The bay is no longer a place in itself.

It is part of something moving through it.

This is where the significance of the whale becomes clear.

Not as spectacle.

Not as interruption.

But as a living expression of continuity.

The whale does not observe Memory Currents.
It lives within them.

Its body is attuned to the same systems that shape the bay, but at a scale and sensitivity that extends far beyond human perception. It responds to what continues. It follows what carries forward.

This is embodied listening in its most complete form.

Not learned.

Inherent.

When you are at the edge of the bay as a whale enters, something shifts in your own perception.

Not because the whale explains anything.
But because it reveals what has always been present.

Attention changes. The body becomes more aware of what it is already sensing.

Subtle movement becomes clearer. Sound carries differently. Space feels altered.

You begin to feel the scale of what you are within.

Not as concept. As experience.

The boundary between what is here and what is beyond here becomes less defined.

The whale does not belong only to this place.
And yet, it is fully within it.

The same is true for the water.

The same is true for you. This is where the relationship between human and whale deepens.

Not symbolic. Not interpretive.

Relational.

You are both within the same system, responding to the same conditions, moving in different ways through the same field of continuity.

The difference is not in whether you are part of it. The difference is in how clearly it is perceived.

The whale does not need to learn how to listen in this way. It is already living within it.

For the human, this is something that must be remembered.

Not as stored knowledge. As reorientation.

When the whale enters the bay, it does not bring something new.

It reveals what has always been moving. And if you are present, if you are staying, if you are listening without taking, you begin to recognise it.

Not as something outside of you. As something you are already within.

Human and Whale

The relationship does not begin with the whale.
It begins with the conditions that hold both.

Water. Movement. Continuity.

The same field that carries the whale carries you.
The difference is not in whether you are part of it.
The difference is in how it is perceived.

For the whale, there is no separation.

It does not step outside of the system to understand it. It moves within it. Responds to it. Continues through it.

Its body is attuned to conditions that extend beyond what can be seen. Temperature, pressure, sound, subtle variations in the water that carry information across distance. These are not signals it decodes.

They are the environment it inhabits.

For the human, this relationship has become less direct. Attention is often placed elsewhere. On surface. On outcome. On interpretation.

The body still responds, but that response is often overlooked.

Diminished. Interrupted.

This is not a failure.

It is a shift in orientation.

At the edge of the bay, this begins to change.

Not because something new is introduced.

Because what is already present becomes more noticeable.

The same wind that moves the surface of the water moves across your skin.

The same light that shifts across the bay shifts across your field of vision.

The same conditions are moving through both.

The difference is in how fully they are registered.

When a whale enters the bay, this difference becomes more visible.

Not because the whale is separate from you.

Because it reveals a way of being within the system that is continuous.

Unbroken.

It does not hesitate.

It does not question whether it belongs.

It moves as part of the field it is within. This is not something to imitate. It is something to recognise.

The human body is not excluded from this system.
It has not been removed from it. It has only learned to place its attention elsewhere.

When attention returns, the relationship becomes apparent again. Not as an idea. As experience.

You begin to feel that what moves the whale is not entirely separate from what moves you. Not in scale. Not in capacity. But in condition.

The same currents that guide movement through the ocean are part of the same system that shapes the bay. The same field that carries the whale carries the water that reaches the shoreline.

And that same field moves through you. This does not make the human and the whale the same. It reveals that they are not separate.

The whale moves within this continuity with a clarity that does not require reflection. The human moves within it with a capacity for awareness that can be reoriented.

This is where the relationship becomes significant.
Not symbolic. Not interpretive.

Relational.

You are not looking at the whale as something outside of you. You are witnessing a form of participation within the same system you are already part of.

The whale does not represent something. It demonstrates something. Continuity. Embodied listening. Movement within a field that does not need to be defined in order to be followed.

At the edge of the bay, this becomes something you begin to recognise. Not immediately. Not through a single encounter.

Through return. Through staying. Through allowing perception to shift. You begin to feel when you are more aligned with what is occurring.

When your attention is less fragmented. When your body is responding more fully to what is present. These are not dramatic changes.

They are subtle. But they are consistent. The more you notice them, the more they become familiar.

The whale does not create this. It reveals it. It shows what it looks like when movement and response are not separated.

When the body is not held apart from the conditions it is within.

This is not something to aspire to. It is something to remember. Not as knowledge. As orientation.

The human capacity for this has not been lost.
It has been layered over. Redirected. But it remains.

At the edge of the bay, in the presence of the whale, that capacity becomes easier to recognise.

Not because the whale gives it to you.

Because it makes visible what is already there. The relationship is not between two separate beings observing one another. It is within a shared field.

A field of movement.
Of continuity.
Of response.

The whale moves through it. You stand within it.

And over time, if you remain, if you listen, if you allow the body to lead, the distance between those positions begins to soften.

Not disappearing. But no longer fixed.

You are still human. The whale is still whale. And yet, something is shared. Not in form. In condition.

In the way both are held within the same movement. This is the relationship. Not something created. Something recognised.

What Cannot Be Seen

Not everything that moves is visible.

At the edge of the bay, this becomes clear over time. What first draws attention is what can be seen: the surface of the water, the movement of light, the shift of wind across the bay. These are immediate. They offer themselves easily.

But they are not the whole of what is occurring. Beneath the surface, movement continues.

Currents move without outline. Temperature layers shift without being marked. Sound travels across distance without leaving a trace that can be followed by sight.

Pressure changes, subtle and constant, shaping what unfolds without announcing themselves.

None of this is hidden. It is simply not visible.

The tendency is to give priority to what can be seen. To trust what presents clearly. To assume that what is not visible is either absent or less significant.

At the bay, this assumption begins to loosen. The longer you remain, the more it becomes apparent that what is visible is only a fraction of what is active.

The surface is an expression. Not the source.

Movement below influences what appears above. Conditions that cannot be seen shape the patterns that can. What is visible is always being informed by what is not.

This is not a separate layer. It is part of the same system. The visible and the invisible are not divided. They are continuous.

The body begins to register this before the mind does.

A shift in awareness that does not correspond to something you can point to. A sense that something has changed, even when nothing obvious has.

These moments are easy to dismiss. They do not come with confirmation. They cannot be easily explained. And yet, they are consistent.

The body is responding to conditions that are not immediately visible.

Subtle changes in air pressure. Variations in sound carried across water. Movements beneath the surface that alter what is felt, even if they are not seen.

This is not unusual.

It is simply not often attended to.

At the edge of the bay, attention begins to widen. Not to include more visible detail. To include what cannot be seen.

This does not mean imagining. It does not mean assigning meaning where none exists.

It means allowing for the possibility that what is being registered is not limited to what is visible.

Listening shifts again here. It is no longer tied to the surface. It extends into the field of what is present but not outlined.

This is where the practice deepens. You begin to trust what is felt, even when it cannot be confirmed through sight.

Not as belief. As recognition.

The same is true in larger systems.

Ocean currents move beneath the surface, shaping climates, carrying heat, influencing conditions across distance. They are not visible from above, and yet their effects are undeniable.

The same is true of migration.

The pathways followed by whales are not marked on the water. They cannot be seen from the shoreline. And yet, they are followed with consistency across generations.

The absence of visibility does not indicate absence of presence. It indicates a different mode of perception is required.

At the bay, this becomes something you begin to accept.

Not everything will show itself clearly. Not everything will be available to sight.

And yet, it is still active. Still moving. Still part of the system.

This changes how you relate to what is in front of you.

You are no longer relying solely on what can be seen. You are allowing for what can be felt. What can be sensed. What can be recognised without being outlined.

This does not replace the visible. It completes it. The surface remains. Movement continues. Light shifts. Wind passes.

But now, these are understood as part of something larger. Expressions of conditions that extend beyond what is immediately apparent.

The bay is not only what you see. It is what continues beneath, within, and beyond that surface.

This is where Listening Lines extend further. They are not confined to what is visible. They move through what cannot be seen.

Felt through the body. Recognised through continuity. Not mapped. Not outlined. But present.

And once this is recognised, the absence of visibility is no longer a limitation. It becomes an opening.

A way of perceiving that is not dependent on what presents itself clearly. A way of being within the system that does not require everything to be known in order to be felt. Nothing has been added. Nothing has been revealed that was not already there.

Only this - You are no longer limited to what you can see.

And in that, the field of what can be perceived expands.

Formed Through Return

Nothing in this work is established in a single moment.

Not the patterns. Not the recognition. Not the shift in perception.

All of it is formed through return.

At first, returning appears simple. You come back to the same place. The same shoreline. The same edge of the bay.

But it is never the same. Conditions change. Light shifts. Wind alters direction.

The surface never repeats itself. What remains is not sameness. It is continuity.

Return is not about revisiting an identical moment.
It is about entering the same field under different conditions.

Each time, something is familiar. Not in form. In orientation.

The body begins to recognise the place before anything is consciously observed. A subtle settling. A shift in breath. A sense that you have entered something that is already in motion.

This recognition is not constructed. It is accumulated.

One visit does not establish it. Two visits begin to suggest it. Over time, it becomes unmistakable.

Not because the place has become predictable. Because you have become more attuned to what continues.

Return allows for this attunement. Without it, everything remains singular. Isolated. Unconnected.

Each moment stands alone, without reference to what has come before. With return, something begins to link. Not through memory as recall. Through memory as continuity.

You begin to sense how one moment relates to another, even when they appear different on the surface.

The way the wind moves today carries something of how it moved before. The way the water responds holds a familiarity that cannot be reduced to repetition.

This is not recognition of exactness. It is recognition of pattern. And pattern can only be felt across time.

Return is what makes time available to perception. Not measured time. Experienced time.

Time that is not divided into segments, but felt as a field through which movement continues.

Without return, there is no depth to perception. Only surface. Only what is immediately present, without context.

With return, depth forms.

Gradually. Quietly. Without announcement.

You begin to notice more. Not because you are looking harder. Because what was previously unnoticed has had time to become familiar.

The body learns this before the mind. It begins to anticipate without predicting. To recognise without naming. To orient without instruction.

This is how Listening Lines form.

Not as something discovered in a single moment. But as something that becomes apparent through repeated presence.

Each return adds something. Not information. Familiarity.

A deepening sense of what continues. What once felt undefined begins to hold shape. Not a fixed shape. A relational one.

You begin to feel where you are within the field.
Not as position. As participation.

Return also changes the way you experience difference.
What once felt entirely new begins to feel connected.
What once appeared separate begins to reveal continuity.

This does not reduce variation. It reveals what holds it together.

The bay does not become predictable. It becomes knowable in a different way.

Not through control. Through relationship.

This is the quiet discipline of return. There is no reward in the conventional sense.

No clear outcome. No moment of completion. Only a deepening.

You come back. You remain. You notice more than you did before.

And over time, without needing to define it, something becomes clear.

What you are perceiving is not a series of separate experiences. It is a continuous movement.

One that you are entering again and again. One that you begin to recognise not because it repeats exactly, but because it continues.

Return is what makes this visible. Not all at once. But gradually.

Until what was once unseen becomes part of how you perceive. And once it is recognised, it does not leave.

Even when you do.

Because what you return to is not only the place. It is the field of continuity that remains in motion.

And through return, you begin to recognise that you are part of that movement.

Not occasionally.

Consistently.

Whether you are aware of it or not.

Entering the Water

There is a point where the edge is no longer enough.

You have stood.
You have watched.
You have stayed.

You have learned to remain without taking, to listen without needing to conclude, to recognise what continues beneath what changes.

And still, there is a boundary. The shoreline holds it. A line between where you stand and where movement carries differently.

At first, that line is sufficient. Observation happens from here. Perception begins from here.

But over time, something shifts. The distance becomes noticeable.

Not as a problem. As a limitation.

You are close to the movement, but not within it. You can see the water respond, but you are not fully responding with it.

The body senses this before the mind does.

A pull. Subtle at first. Not an instruction. An inclination.

To step forward. Entering the water is not an act of courage.

It is an extension of listening.

There is no ceremony required.

No preparation beyond what has already taken place through staying and returning.

You step in. The temperature registers immediately. Not as information. As sensation.

The body adjusts. Breath changes. The skin responds. Everything becomes more direct.

The distance that existed at the shoreline dissolves. You are no longer observing the conditions. You are within them.

The movement of the water is no longer something you see. It is something you feel. Around your legs. Against your body.

In the way balance shifts with each small change. The same currents that moved across the surface now move through you.

Not metaphorically. Physically.

The body responds without needing to be told how.

Muscles adjust. Posture shifts. Breath aligns with movement in ways that are not planned.

This is not learning. It is recognition.

The body already knows how to be in water. What changes is your awareness of that knowing.

Sound alters. What was once carried through air now moves differently. More contained. More immediate. Less defined, but more present.

Light behaves differently. Refraction shifts the way you see. The surface is no longer something you look across. It is something you are part of.

There is no stable point. No fixed position.

Everything moves. And you move with it.

Not by choice. By condition.

This is where embodied listening deepens again.

At the shoreline, you began to sense the continuity of movement. In the water, you participate in it.

The distinction between observer and environment reduces further. You are no longer outside the system. You are within the same dynamics that shape it.

Temperature, pressure, movement, sound. All experienced directly. All influencing response.

This is not immersion as activity. It is immersion as relationship.

Nothing needs to be achieved. Nothing needs to be held.

Only this – Remain, Feel and Allow the body to respond.

There is a moment, often brief, where the need to control returns. To stabilise. To orient in a familiar way.

This does pass. The body adjusts. Not to stillness, to movement.

You begin to trust the instability. Not as uncertainty. As continuity. The water is never still. And yet, it is not chaotic. It carries itself.

You begin to feel that carrying. Not as something separate from you. As something you are within.

Time changes again here. Not measured. Not tracked. Experienced through sensation.

Moments stretch without needing to be counted. Attention does not fix. It moves. With the water. With the body. With what continues.

This is where Listening Lines shift again.

They are no longer something you sense from the edge. They move through you.

Felt in the way the body aligns with movement. In the way awareness adjusts without instruction. In the way continuity becomes something you participate in, not just recognise.

There is no clear point where this completes. No moment of arrival. Only a deepening.

You remain in the water. You feel more. You respond more fully. And gradually, without needing to define it, something becomes clear. You are not entering the system. You were always part of it.

Stepping into the water does not create that relationship. It removes the distance that made it harder to feel. And once that distance is gone, even briefly, it does not fully return.

You step back onto the shoreline. The body carries what it has registered. Not as memory stored. As orientation.

A subtle shift in how you stand. How you listen. How you perceive what is already moving.

Nothing has been added. Nothing has been taken.

Only this - You have felt the system from within. And that cannot be unseen. Even when you return to the edge.

Lines Felt Through the Body

After entering the water, something changes in how the lines are perceived.

They are no longer approached from the outside.
They are felt from within.

At the edge of the bay, Listening Lines begin as something subtle. A sense of continuity. A recognition that movement carries forward beyond what is immediately visible.

In the water, that recognition becomes more direct.

The body begins to register alignment. Not as a concept. As sensation.

A slight ease in one direction. A resistance in another. A shift in balance that does not come from conscious decision.

These are not instructions. They are responses.

The body is orienting within a field of movement. The lines are not fixed paths. They are not routes that can be followed or mapped. They are patterns of alignment within what is already moving.

Felt through the way the body responds. At first, this can feel uncertain. There is no visual confirmation. Nothing to point to and say, this is it.

The mind looks for clarity. For something defined.

But the body does not work in that way. It does not require fixed reference points. It responds to relationship.

The way water moves around you informs how you stand. The way currents shift informs how you adjust. These are small movements. Often unnoticed. And yet, they are consistent.

The more attention is given to them, the more they become apparent. Not because they become stronger. Because they are no longer overlooked.

You begin to sense when you are moving with the water. And when you are moving against it. Not as effort. As alignment.

There is a difference.

When aligned, movement feels continuous. Supported. Carried.

When not aligned, there is friction. Not discomfort. A subtle sense of interruption. This is not something that needs to be corrected. It is something that can be felt.

The body recognises the difference without needing to name it.

This is where Listening Lines become more distinct. Not as visible lines. As pathways of least resistance within a moving system. Not fixed. Not permanent. Always adjusting. Always responding to the conditions present.

You do not find them. You feel when you are within them. This is not exclusive to the water. It begins there. It becomes noticeable there. But it does not end there.

The same principle applies at the shoreline. On land. In movement.

The body is constantly orienting. Responding to conditions that are not always consciously recognised. What changes is your awareness of it.

Once felt in the water, it becomes easier to recognise elsewhere. A sense of when something flows. When something continues without interruption. When movement feels supported rather than forced.

This is not abstract. It is physical. It is immediate. It is felt.

The lines are not separate from the system. They are expressions of it.

They exist because movement exists. Because continuity exists. Because conditions are always interacting.

The body is part of that interaction. Not observing it. Participating within it.

This is why the lines can be felt. Not because they are marked. Because the body is already responding to the same conditions that create them.

This does not require skill. It does not require training. It requires attention. Not focused in a narrow way.

Open. Responsive.

Allowing sensation to inform perception. The more this is allowed, the more consistent it becomes. Not in form. In recognition.

You begin to trust what is felt before it is confirmed. Not as belief. As experience.

There is no need to define the lines. No need to locate them precisely. The moment they are fixed, they are no longer what they are.

They are movement. They are response.

They are continuity expressed through the body. This is where Listening Lines fully shift. From something you are trying to perceive. To something you are already within.

Felt.

Not seen. Not held. Not followed. But recognised.

In the way the body aligns with what continues. And once recognised, even briefly, this changes how you relate to movement.

Not only in the water. Everywhere.

Because the system does not stop at the shoreline. And neither does your participation in it.

The lines do not begin where you step in. And they do not end when you step out. They are present. Always.

The only question is whether they are being noticed. And once they are felt through the body, they do not need to be searched for again.

They are already there. Waiting to be recognised in the way you move, the way you stand, the way you respond.

Not separate from you. Part of the same continuity you have been entering all along.

Returning Again and Again

There is no single moment that establishes this.

No point at which you arrive and everything becomes clear.

What forms here is not built through intensity.
It is built through repetition.

You come back. Not because something dramatic has happened. Because something quiet has begun to hold.

At first, returning can feel unnecessary. You have already been here. You have already seen the water, felt the wind, noticed the movement.

There is a sense that you understand it. That you have taken in what is there.

But this work does not reveal itself in a single encounter. It does not offer itself fully in one moment of attention.

Each return is different. Not because the place has changed entirely. Because the conditions have shifted.

Light falls differently. Wind moves differently. The water carries itself in a new way. Nothing repeats exactly.

And yet, something continues. This is what begins to draw you back. Not curiosity. Recognition.

A sense that what you are entering is already in motion. That you are not starting again. You are re-entering.

This distinction matters.

Starting again suggests something separate. A new beginning. Return suggests continuity. An ongoing movement that you step back into.

Each time you arrive, the body recognises before the mind does. A subtle settling. A familiarity that cannot be traced to a specific memory.

You do not need to orient yourself in the same way. The place begins to orient you. This is how depth forms. Not through effort. Through repetition.

The more you return, the less each visit stands alone.

Moments begin to connect. Not in sequence. In relationship. What you notice today carries something of what you noticed before.

Not as a comparison. As continuity.

You begin to feel how one condition influences another. How movement carries forward. How response is shaped by what has already occurred.

This cannot be understood from a distance. It must be experienced across time.

Return is what allows time to become perceptible in this way. Not measured. Felt.

Time begins to lose its edges. It is no longer divided into distinct moments. It becomes a field through which movement continues.

Each return deepens this - not dramatically, gradually. Quietly. Without announcement.

You notice more. Not because you are looking harder. Because what was once unfamiliar has become known in a different way. Not named. Recognised.

The body holds this recognition. Not as stored information. As orientation.

You begin to sense when you are within the continuity of movement. And when you are not.

This is not a judgment. It is a sensitivity. A refinement of perception that comes only through repeated contact.

There is no shortcut to this. No way to accelerate it. It does not respond to intensity. It responds to consistency.

You come back. You remain. You allow what is already moving to become more apparent.

Over time, the need for confirmation falls away. You do not require a clear signal that something has happened.

You begin to trust the subtle. The barely noticeable. The quiet shifts that do not announce themselves but remain consistent.

This is where the practice becomes less defined. Less something you are doing. More something you are within.

Return is no longer an action. It becomes a rhythm. Something that carries itself.

You find yourself drawn back without needing a reason. Not out of habit. Out of alignment.

A sense that this is where you can feel what continues.

The bay does not ask you to return. It continues whether you are there or not. But through return, your perception of that continuity deepens.

You begin to understand that what you are returning to is not the place alone. It is the movement. The field. The ongoing interaction that does not pause in your absence.

Each time you step back into it, you are not beginning again. You are continuing.

This changes how you experience presence. You are no longer trying to capture a moment. You are entering something that is already unfolding.

And in that, something settles. There is less urgency. Less need to define what is happening.

More capacity to remain within it. Return reveals that nothing is lost between visits. Everything continues.

You simply become more able to recognise it. And once that recognition begins, it does not reset.

It carries forward with you. Into the next return. And the next. And the next.

Until what once felt like separate visits becomes something else. A continuous relationship. Not dependent on time spent. But on continuity recognised.

You are no longer visiting the bay. You are part of the movement you keep returning to.

And through that, return is no longer something you decide to do.

It is something that is already happening.
You are simply stepping back into it, again and again.

The Discipline of Attention

Attention, as it is usually practised, moves quickly. It scans. Selects. Moves on.

It looks for what stands out, what can be understood, what can be used. It is shaped by purpose. Directed toward outcome.

At the edge of the bay, this way of attending begins to feel insufficient. Not because it is wrong. Because it cannot hold what is occurring here.

What continues at the bay does not present itself as something to be captured quickly. It does not offer a clear focal point. It does not reward speed.

It requires something else.

Attention that remains. This is where discipline begins. Not as control. Not as effort applied against resistance. As consistency.

A willingness to stay with what does not immediately resolve into meaning.

At first, this can feel like doing nothing. There is no clear task. No instruction beyond remaining.

The mind looks for something to engage with. Something to fix on. Something to define.

Without that, attention begins to drift. It moves away from the water. Toward thought. Toward what is next.

This is the habit of attention. It does not stay unless given a reason. The discipline is not to force it back. It is to notice when it has moved.

And to return. Again, and again.

Without frustration. Without judgment.

This is simple. And not easy.

Because nothing is being produced. Nothing is being achieved in the way attention is usually measured.

And yet, something is forming.

Attention begins to change its pace. It no longer needs to move quickly. It no longer needs to select a single point of focus. It widens. It holds more than one thing at once.

The movement of water. The sound carried across it. The feeling of wind against the body.

All present.

None needing to be prioritised. This is not distraction. It is expansion.

Attention becomes less narrow, less driven by the need to isolate and define. It becomes more receptive. More able to remain with what is unfolding without needing to reduce it.

This is where the discipline deepens. You are not holding attention on something. You are allowing attention to remain open.

This requires a different kind of steadiness. Not rigid. Responsive. The ability to stay with what is changing without needing to control it.

At the bay, this becomes possible because there is nothing asking to be completed. No endpoint. No resolution. Only continuation. Attention begins to align with that. It moves less like a tool and more like a field.

Something that can hold what is present without needing to organise it immediately. This does not remove thought. It changes its role.

Thoughts no longer leads. They follow. It reflects after perception has already occurred.

This is where clarity begins to form. Not through effort. Through alignment.

Attention is no longer pulling experience apart. It is allowing it to remain whole.

This is what makes deeper perception possible. Not because more information is gathered. Because less is lost.

The discipline is not about intensity. It is about return.

Each time attention moves away, you notice. You return.

No force. No correction. Just a quiet reorientation.

Over time, this becomes more natural. Attention settles more quickly. It remains for longer. It becomes less reactive to distraction. Not because distraction has disappeared. Because it no longer pulls attention in the same way.

This is not something that happens all at once. It forms gradually. Through repetition. Through consistency. Through the same quiet practice that underlies everything in this work.

You come back. You remain. You notice when attention has moved. You return it. Again, and again.

Until the movement of attention itself begins to change. It becomes less restless. Less driven by the need to find something. More able to be with what is already present.

At this point, attention is no longer something you apply. It becomes part of how you are within the system. It aligns with what continues. It moves with the same steadiness.

The same responsiveness. The same lack of urgency. This is the discipline. Not imposed. Formed.

Through repeated contact with something that does not respond to force. Only to presence.

And once this is established, even lightly, it carries forward. Into other places. Other moments. Other conditions.

Because the discipline is not tied to the bay. It is something that has been formed through it.

A way of attending that does not rely on outcome. That does not collapse when nothing is happening. That remains.

And in remaining, allows what is already present to become perceptible. Not all at once. But steadily.

Continuously.

As long as attention is willing to stay.

The Women Who Enter the Water

It is not only an individual practice.

At the edge of the bay, what begins quietly in one body is also happening in others.

Not organised. Not instructed. Not defined as a shared method.

And yet, shared.

There are women who enter the water regularly. Not occasionally. Not as an event. As rhythm.

Across the shoreline of Williamstown, they gather in the early light, in the shifting weather, in conditions that do not announce themselves as ideal. They come at different times, in different groupings, with different reasons for arriving.

And yet, what they are doing is consistent.

They enter the water.

Some slip beneath the surface with masks and fins, moving slowly through the shallows of Jawbone Marine Sanctuary, where reef, seagrass, and rock hold a different kind of stillness.

Others do the same further along the bay at Ricketts Point Marine Sanctuary, where the water clears just enough to reveal what cannot be seen from above.

They are not separate practices. They are extensions of the same movement.

To enter. To remain. To be within.

Whether standing at the shoreline, swimming through open water, or drifting slowly above reef, the orientation is the same.

The body is placed into the water and allowed to respond. There is no performance in this. No need to explain.

They move from land into water as part of a rhythm that does not need to be named in order to be followed.

Some come together. Some arrive alone and meet others already there.

Some enter quietly, without conversation.

The form varies. The movement does not. They step in. They feel the temperature. They adjust. They remain.

Among those who swim strokes, snorkel and dive, there is an added layer of immersion.

The surface is no longer the primary reference.

They move their faces beneath it, allowing vision to reorganise, allowing breath to regulate through rhythm, allowing sound to change.

The world above softens.

The world below becomes present.

Fish move without urgency.

Light filters differently.

The body floats within a field that is quieter, more contained, yet still in motion.

This is not observation from above. It is participation within.

The same principle holds. Not taking. Not extracting. Remaining.

Across the bay, this continues.

Different groups. Different locations. Different conditions. And yet, the same act.

Women entering the water. Returning. Again, and again.

This is not confined to one place. It is not confined to one group. It is not organised.

And yet, it is happening. Quietly. Consistently. Without needing to be named.

Men enter the water too. They swim, they dive, they snorkel, they move within the same conditions and are part of the same system.

This is not a distinction of exclusion. It is a distinction of emphasis.

This work is paying attention to something specific. The relationship between water and the feminine. Not as gender alone. As energy. As origin. As continuity.

Water is the first environment of human life. The body forms within it. Held, sustained, shaped.

The fluids of the body carry this forward. Salinity, rhythm, flow. The same elemental qualities that exist in the ocean exist within the body.

This is not symbolic. It is physical. And it is also relational.

The feminine, in this context, is not defined by role or identity. It is defined by capacity.

To hold. To carry. To allow life to form and continue. Water expresses this.

And many women, consciously or not, are responding to it. Returning to it. Entering it. Allowing the body to reconnect with a condition it already knows.

This is not something they are being taught. It is something they are recognising.

The women who enter the water are not performing a concept. They are participating in a relationship.

One that exists whether it is named or not. One that continues whether it is explained or not.

Across the bay, this is visible. In the early morning swimmers. In the quiet groups at the shoreline. In the snorkellers moving slowly across reef. In the solitary figure entering the water without announcement.

Different expressions. The same continuity.

This is where the work extends beyond the individual. It becomes collective. Not through coordination. Through shared orientation.

Each woman entering in her own way. And yet, all within the same field. The same movement. The same continuity.

This does not require agreement. It does not require language.

It is enough that it is happening. That the water is being entered. That the body is being allowed to respond. That the relationship is being maintained. Quietly. Repeatedly. Without needing to be named.

This is where Listening Lines expand again. They are not only felt individually. They are being lived across bodies.

Across places. Across the bay. Not as something organised. As something remembered.

The women who enter the water are part of this continuity. Not separate from it. Not observing it. They are within it.

And in that participation, something becomes clear. This is not new. It has not been created here. It is being returned to.

Across the bay. Across time. Across bodies that continue to enter, again and again, into what has always been moving.

And in that, something is held. Not formally. Not visibly. But unmistakably.

A relationship between water and life. Between body and movement. Between the feminine and the field that carries it.

Not defined. Not contained. Only continued.

The Crystals (Rockpools at Williamstown Beach)

She arrives at the steps.
The rail runs down the centre, worn by hands that have come before her.
She takes it lightly, not needing to hold, just to meet it.
The water begins at the foot of the stairs.
No shoreline.
No gradual entry.
Contact is immediate.
She pauses.
The temperature moves through her quickly.
Feet. Ankles. Calves.
A small laugh escapes.
Not from humour.
From recognition.
The body adjusting.
The water being real.
She steps down again.
Hand on rail.
Weight shifting.
Then off the final step and onto the rocks.
Uneven.
Alive beneath her.
At low tide, they stretch outward, exposed, waiting.
She feels for footing.
Places herself carefully, then more fully.
Her hands move through the water.
Back and forth.
Acclimatising.
Reintroducing.
The noodle rests beside her.
Floating. Waiting.
She moves forward.

Not far.
Just enough.
Then she pushes gently out and over the rocks.
And the body is no longer standing.
It is held.
Carried.
Below her, the sea forest begins.
Not in the distance.
Here.
Movement already underway.
Seaweed rising and falling.
Life moving through it without interruption.
She leans into the water.
Allows it to take her weight.
Small movements begin.
Legs extending.
Arms moving slowly.
Not exercise.
Alignment.
Breath settles.
Inhale.
Exhale.
The body finds its place.
Nothing to reach.
Nothing to complete.
She remains.
And when she returns,
hand to rail,
foot to step,
body lifting back into air,
something has shifted.
Not added.
Not taken.
Just known.

Below the Surface, Together with Jen1

We did not arrive as experts.
We arrived with curiosity,
with breath to learn,
with a willingness to enter something we did not yet understand.
Jen1 knew the way the shoreline changes
once you stop standing
and begin to float.
She did not explain it in full.
She didn't need to.
"Just come Jen2," she would say,
already in the water,
already part of it.

At Jawbone Marine Sanctuary
the surface gives very little away.
From above, it is wind,
tide,
movement without detail.
But beneath -
it opens.

I wore fins.
Discovering distance.
Discovering tide.
Discovering movement in this body of water.
Jen1 did not need them.
She has visited many times before,
as though the body already knew
what mine was still figuring out.

With body stretched more horizontal than upright,
the world rearranges.

Not reefs -
but forests.
Not structure -
but gardens.
Living strands rising and falling,
swaying in rhythm that does not belong to the wind.
Growth layered upon growth.
Movement within stillness.

At first,
I did not know what I was looking at.
Only that it was more
than I expected.

Jen1 gestures,
not urgently,
not to direct -
but to include.
There.
And in that small act,
something shifted.
Not just what I saw,
but how I saw.

Fish move through the sea gardens,
darting,
pausing,
holding themselves in the current.

The under-water forests seem endless -
no beginning,

no end.
Only movement.

There were days of clarity
and days of distance.
Days when everything came close,
and days when it held itself just beyond reach.
None of it lost.
All of it part of learning
how to be there.

We surfaced sometimes and spoke.
Often we didn't.
The words were never the point.
What mattered was
that we returned.
Again.
And again.

Jen1 showed me that the Sanctuary
is not what it first appears to be.
That what looks empty
is often full.
That what looks simple
holds entire systems of life beneath it.

Long before the whales recently visited these waters,
before questions stretched outward,
before meaning asked more of me -
this is where it began.

Face in the water.
Breath steady.
Body adjusting.

Not trying to master it.
Just learning how to stay.

Jen1 and Jen2.
One guiding.
One learning.
Both returning.

Not separate from the place.
Not claiming it.
Just present long enough
for something to reveal itself.
And to realise -
it had been there the whole time.

Jawbone by One

She places her towel on the rock.
Ancient.
Worn smooth by time and tide.
It holds without needing to be noticed.
She steps toward the water.
The sand begins first.
Bare.
Open.
Unassuming.
Each step sinks slightly beneath her.
The ground shifts.
The water moves around her legs.
There is time here.
Time to enter.
Time to feel.
No sudden drop.
No immediate immersion.
Just a gradual meeting.
She walks further.
Then lowers herself.
The mask comes down.
The breath changes.
The fins go on.
And the transformation begins.
She leans forward.
The body turns from vertical to horizontal.
Held at the surface.
Breathing continues.
Steady.
Unbroken.
Through the snorkel.
She does not disappear beneath the water.

She hovers above it.
Like an aircraft suspended in air.
Moving slowly across what is already alive below.
What was not visible from above becomes immediate.
An underwater forest.
Not still.
Alive.
Seaweed rising and falling in slow rhythm.
Fish moving through it without urgency.
Light filtering down, shifting across everything without breaking what continues.
The surface holds her.
The water carries her.
There is no need to dive.
No need to go further.
Everything is already here.
She moves slowly.
Or not at all.
The water moves her.
The breath sets the rhythm.
Inhale.
Exhale.
Continuous.
Uninterrupted.
The body remains at the threshold.
Between air and water.
Between above and below.
Not separate from either.
The boundary softens.
Not gone.
Less fixed.
She remains.

And when she lifts her head slightly, clears the snorkel,
and turns back toward the shore, stepping again across the
sand that once appeared empty,
she knows it is not.
It never was.

Pope's Eye

She arrives where the bay opens.
Not contained.
Not held in the same way.
The water moves differently here.
More ocean than shoreline.
At Pope's Eye, the structure rises from the water, man-made, but settled into place long enough to feel part of what surrounds it.
Inside, the water is shallow.
Two, maybe three metres.
Contained.
Held within the ring.
Outside, it drops.
Not dramatically.
But enough.
Eight. Ten metres.
A shift.
The gear is prepared.
Tank.
Weights.
Mask.
Fins.
Each piece added deliberately.
Not to go further.
To change how the body meets the water.
She steps in.
No gradual entry.
No time to adjust in stages.
The water takes her immediately.
The breath changes.
Drawn through the regulator now.

Audible.
Measured.
Inhale.
Exhale.
Slower.
More deliberate.
She descends.
Not deep.
Not far.
But enough to feel the difference.
The surface is no longer holding her.
She is within the water.
Suspended.
Not floating above.
Not standing below.
Held in between.
The structure of the reef moves differently here.
Closer.
More present.
Not distant as it is from the surface.
The same marine life.
The same movement.
But encountered differently.
The body is not observing from above.
It is within the same density.
The same pressure.
The same field.
Nothing dramatic changes.
And yet, everything shifts.
The breath becomes the rhythm.
Each inhale lifts slightly.
Each exhale releases.
Buoyancy is no longer assumed.
It is felt.

Adjusted.
Aligned.
There is no need to go deeper.
The difference is not depth.
It is relationship.
To be within, rather than above.
To move with, rather than over.
She remains.
As long as the breath allows.
And when she rises again, slowly, returning to the surface, the boundary re-forms.
Air.
Light.
Distance.
But something stays.
Not as image.
As knowing.
That the bay is not only what is seen from above.

Chinaman's Hat

At Chinaman's Hat, the seals have a place.
A structure rising from the water.
A haul-out.
A home base.
They return here.
Again, and again.
Resting.
Gathering.
Not passing through.
Holding.
You enter the water nearby.
Not to approach.
Not to interrupt.
Just to be within the same field.
They notice.
Before you do.
A head lifts.
Then another.
Eyes on you.
Not alarmed.
Aware.
Then movement.
One slips in.
Then another.
The water changes.
Not in surface.
In presence.
They move around you.
Fast.
Effortless.
Curious.
They circle.

Return.
Disappear and reappear without warning.
There is no straight line in their movement.
No fixed direction.
Only play.
Only ease.
They are completely at home here.
The body feels this immediately.
The difference.
You are within the water.
They are of it.
The response is not to chase.
Not to reach.
Only to remain.
To hold your place within the movement.
They come close.
Closer than expected.
Then gone again.
Then back.
Nothing asked.
Nothing given.
And yet, something shared.
A moment of being within the same conditions.
Without needing to define it.
They return to the structure.
Climb out.
Rest again.
The pattern continues.
You leave the water.
They remain.
This is their place.
And for a moment, you were within it.

Dolphins of the Bay

Not beheld to one place.
No structure.
No fixed point.
They move.
Across the bay.
At times near Williamstown, at times further south, at times not at all.
You do not go to them.
You encounter them.
When they arrive, it is immediate.
The field shifts.
Before they are seen.
Then movement.
Fast.
Precise.
Through the water.
They do not circle in the same way as the seals.
They pass.
Return.
Pass again.
A different kind of interaction.
Less contained.
More directional.
They are not anchored here.
They are moving through.
You are in the water when they come.
Not waiting.
Not expecting.
Just present.
And suddenly, they are there.
The body responds.
Not through thought.

Through awareness.
Speed.
Proximity.
A different rhythm in the water.
They do not slow for you.
They do not need to.
You are part of the field they are moving through.
For a moment.
Then they are gone.
No clear beginning.
No clear end.
Only passage.
The water settles.
But not completely.
Something remains.
Not as memory held.
As imprint.
As a recognition of movement that extends beyond what can be seen.
They are everywhere.
And nowhere fixed.
Part of the same continuity.
Never held.
Always moving.

Holding the Edge

There are times when you do not enter.

Not because you cannot. Not because you are unsure. Because you choose to remain.

At the edge of the bay, this becomes its own practice. Not lesser than entering. Not a step before something else.

Complete in itself.

You stand where land meets water, where what holds meets what moves. It appears as a boundary, but it is not fixed. It shifts. It responds. It is held in relationship.

The water does not stop at the edge. It continues beneath.

It carries movement beyond what can be seen. The land does not remain unchanged either. It gives, it erodes, it adjusts over time.

The edge is not a line. It is a meeting.

To stand here is to feel both at once. The stability of ground. The movement of water. Not alternating. Simultaneous.

The body registers this without needing to think about it. One part of you held. Another part adjusting. Even in stillness, there is movement.

This is where attention changes again. There is nothing to reach for. Nothing to complete. Nothing that requires you to go further.

Only to remain.

The edge is often treated as a place to pass through. A transition. A brief pause before entering or leaving.

Rarely as a place to stay.

And yet, it holds something specific.
A form of perception that does not depend on immersion.
You can see the surface. Track the movement. Notice how the wind moves across the water, how the tide shifts, how patterns form and dissolve.
You can feel the air, the temperature, the subtle changes that move through the body as the conditions shift.
You are not separate from what is happening.
You are within it.
Just positioned differently.
This is where restraint becomes clear.
Not holding back.
Not avoiding.
Choosing not to move further.
Even when you could.
This is not hesitation.
It is awareness.
A recognition that depth is not only found by going udeeper.
It is found by remaining.
Allowing what is present to become more fully perceived without needing to change your position within it.
The edge offers clarity.

Less immersive.
More defined.
You can see the movement without being carried by it.
You can feel continuity without dissolving into it.
This is not better than entering the water.
It is not less.
It is another way of being within the same field of movement.
At times, this is what is required.
To stand.
To remain.
To feel what continues without needing to move further.
There is nothing missed here.
What moves does not require you to enter it in order for it to continue.
It is already carrying forward.
The edge allows you to recognise that without needing to step into it.
You begin to understand that proximity is not the only way to deepen perception.
That closeness is not always physical.
That sometimes, the most precise place to be is exactly where you are.
The edge becomes a position.
Not a transition.
A way of being that holds both movement and stillness together.
You are not drawn forward.
You are not pulled back.
You remain.
This requires the same discipline.
The same attention.
The same willingness to stay without outcome.

To allow what is present to continue without needing to follow it.
You stand.
You feel the ground.
You feel the water moving nearby.
You notice what continues.
Not through immersion.
Through presence.
This is where another form of Listening Line becomes apparent.
Not one felt through the body moving within water.
One recognised through the relationship between what holds and what moves.
The line is not fixed.
It shifts.
It responds.
And yet, it is always there.
The edge reveals this.
Quietly.
Without instruction.
You begin to recognise when to enter.
And when to remain.
Not as a rule.
As a response.
To conditions.
To the body.
To what is present.
There is no single way.
Only alignment.
At times, you move into the water.
At times, you stay at the edge.
At times, you move between the two.
Both belong.
Both reveal different aspects of what continues.

The edge holds this balance.
Without needing to resolve it.
Without needing to choose one over the other.
Only to recognise what is here.
And to remain, when that is what is required.
Not as limitation.
As clarity.
You are not separate from what is moving.
Even when you stand still.
Even when you do not enter.
You are still within what continues.
Still part of the movement.
Still responding.
Still listening.
The edge does not remove you from what is happening.
It reveals that you were never outside of it.
Only positioned differently within it.
And that, too, is part of the practice.

Recognition Without Proof

There is a point where what you are noticing cannot be confirmed in the usual way.

No measurement. No clear demonstration. No single piece of evidence that can be pointed to and held up as proof.

And yet, it is not uncertain.

At the edge of the bay, this becomes familiar.
You begin to recognise patterns that are not fixed.
You feel shifts that are not visible.
You sense continuity that cannot be isolated into a single moment.

None of this arrives as something you can prove. It arrives as something you know.

Not as belief. Not as assumption. As recognition.

This kind of knowing is often dismissed. Because it does not present in a way that can be easily explained. It cannot be shown in a single instance. It does not resolve into something that can be repeated exactly on demand.
And yet, it is consistent.

It appears again and again. Through return. Through attention. Through the body registering what continues even as everything appears to change.

This is where a distinction becomes important.

Between proof and recognition. Proof requires something to be fixed. Something that can be isolated, measured, repeated in the same way.

Recognition does not. Recognition forms through continuity. Through exposure over time. Through the body learning to register what persists across variation.

At the bay, this is how perception deepens. You do not prove that movement continues. You recognise it.

You do not prove that patterns carry forward. You feel how they do.

You do not prove that what has been, still influences what is. You notice that it does.

This is not a rejection of science. It is a different mode of knowing. One that exists alongside measurement, not in opposition to it.

Science can describe currents, temperature shifts, migration pathways, the behaviour of living systems. It can map. It can model. It can explain. And this is essential.

But there is also the experience of being within those same conditions.

Of sensing what cannot be reduced to a single variable. Of recognising continuity without needing to isolate it.

These are not competing ways of knowing. They are different perspectives on the same movement.

At the edge of the bay, both are present. The measurable. And the felt. The observable. And the recognised.

You begin to understand that not everything needs to be proven in order to be real. That some forms of knowing arise through direct experience.

Through the body. Through attention sustained over time. This does not mean accepting everything without question.

It means recognising that not all forms of understanding arrive through the same process. Recognition without proof is not vague.

It is precise in a different way. It does not rely on external validation. It relies on consistency of experience.

You notice something once. You question it. You notice it again. You begin to pay attention.

Over time, it becomes unmistakable. Not because it has been proven. Because it has been recognised repeatedly.

This is how Listening Lines become clear. Not through a single moment of discovery. Through accumulation. Through return. Through the body learning to trust what it registers.

This trust is not blind. It is built. Slowly. Through consistent contact with what continues.

At first, you may doubt it. You may question whether what you are sensing is real. Whether it is imagined. Whether it can be relied upon.

This is part of the process. The mind seeks confirmation. The body continues to register.

Over time, the gap between them reduces. What is felt becomes more stable. More recognisable. Less dependent on validation.

You begin to trust the continuity of your own perception. Not because it has been proven. Because it has been experienced consistently.

This is where the practice settles. You are no longer trying to confirm what you notice. You are allowing it to be. You remain open. Attentive. Responsive.

And what continues becomes more and more apparent. Not as something separate from you. As something you are already within.

Recognition without proof does not remove the need for inquiry. It deepens it. You remain curious. You remain attentive.

But you are no longer dependent on a single form of validation. You understand that some things are known through measurement. And some are known through presence.

Both have their place. Both reveal different aspects of what is.

At the bay, this becomes clear. Not as a conclusion. As an ongoing experience.

You stand. You listen. You feel what continues. And over time, you no longer need to prove it in order to recognise it.

It is already known. In the way you perceive. In the way you respond. In the way you remain within what is moving.

This is not certainty. It is something quieter.

More stable.

A knowing that does not need to be held tightly. Because it continues whether you define it or not. And in that, something settles.

You are no longer searching for proof. You are recognising what is already present.

Again, and again.

Each time you return.

Living Within Memory Currents

There is a point where this is no longer something you come to do.

It is something you are already within.

Not only at the edge of the bay. Not only in the water. Not only in moments set aside for attention.

Everywhere.

What began as a practice - to stay, to listen, to notice what continues - begins to move beyond the place where it was first recognised.

It carries.

Not as something you hold. As something that moves with you.

You begin to notice it in other places. The way air shifts before weather changes.

The way a room feels different before anything is spoken. The way the body responds before the mind understands why.

These are not separate from what was felt at the bay. They are expressions of the same continuity.

The same movement. The same capacity to recognise what is already unfolding.

You are not applying what you learned. You are recognising that it was never confined to one place.

The bay made it visible. It did not create it.

This is where something shifts. The idea of "going somewhere to practise" begins to soften.

Not disappear. But expand.

You still return. You still stand at the edge. You still enter the water. But you are no longer limited to those moments.

What you felt there begins to appear elsewhere. Not in the same form. But in the same quality.

A sensitivity to what is changing. A recognition of what continues. A capacity to remain without needing immediate resolution.

This is not something added to your life. It is something that begins to shape how you move within it.

You listen differently. You respond differently. Not dramatically. Subtly. Consistently.

You begin to notice when something aligns. When something carries. When something holds continuity beyond the surface.

And you begin to notice when it does not.

Not as judgment. As awareness.

This changes how you make decisions. How you relate to others. How you move through situations that once felt separate from one another.

You begin to feel the connection between them. Not as an idea. As a condition.

Everything is part of movement. Everything is influenced by what has come before. Everything carries forward in some way.

This does not need to be analysed constantly. It is simply recognised. And in that recognition, something becomes simpler.

You are not trying to control what is happening. You are responding to it. You are not trying to fix everything into place. You are allowing it to move. While staying aware of how you are within it.

This is where the distinction between observer and participant dissolves further. Not completely. But enough.

You are no longer standing outside, looking in. Even when you appear to be. You are within what is happening. Always.

Your body is responding. Your attention is moving. Your presence is influencing and being influenced at the same time.

This is not something you switch on. It is already active.

What changes is your awareness of it. **You begin to trust that awareness**. Not as something perfect. But as something responsive. Something that can adjust as conditions change. Something that does not need to be fixed in order to be reliable.

This is where living within it becomes clear. You are not carrying the practice with you. You are recognising that the practice is simply a way of becoming aware of what is already occurring.

The bay revealed it. The water made it felt. The body recognised it. And now, it continues.

Not because you are maintaining it. Because it is always moving. Always present. Always available to be noticed.

You do not need to hold onto it. You do not need to protect it. You do not need to return to it in a way that separates it from the rest of your life.

You are already within it. In every place. In every moment. To different degrees. With different levels of awareness. But always.

This does not make everything clear.
It does not remove uncertainty.
It does not resolve every question.
It changes how you are within those questions.
You are less likely to force an answer.
More able to remain with what is unfolding.

To recognise that not everything needs to be resolved immediately. That some things continue. And will reveal themselves over time.

This is what the bay has shown. Not something to take away. Something to live within. Not separate from you. Not something you visit.

Something that is already part of how life moves. And in recognising this, something settles again.

You are no longer moving between practice and life. They are the same.

What you felt at the edge of the bay is not confined to that place. It is present everywhere.

You are simply more able to notice it. To respond to it. To remain within it. Without needing to define it. Without needing to hold it.

Only to recognise it. As it continues. Through everything, including you.

A Different Kind of Knowing

There is a way of knowing that does not arrive all at once.

It does not declare itself. It does not resolve into certainty. It does not present as something you can hold and explain completely. And yet, it is steady.

At the edge of the bay, this way of knowing begins to form without announcement. Not through instruction. Not through effort. Through exposure. Through return. Through the body learning to recognise what continues.

At first, it is easy to overlook. It does not feel like knowledge in the usual sense.

There is no clear statement. No conclusion. No moment where you can say, now I understand. Instead, there are small recognitions. A familiarity that builds without needing to be named. A sense that something is consistent, even when it appears to change.

This is where knowing begins differently. Not constructed. Accumulated. Not assembled through thought. Formed through experience.

The mind looks for clarity.

For something defined. For something that can be explained and shared in a fixed way.

This kind of knowing does not offer that immediately.

It remains open. Responsive.

It changes as conditions change. It deepens as attention deepens. It cannot be separated from the context in which it is formed.

This is what makes it different.

It is not abstract. It is relational.

You do not hold it outside of what you are experiencing. It exists within that experience.

At the bay, this becomes more apparent over time. You begin to recognise patterns without needing to identify them precisely.

You feel when something aligns, even when you cannot explain why. You sense continuity without needing to trace it back to a specific point. This is not vague.

It is precise in a way that does not rely on definition. The precision comes from repetition. From consistency. From the same recognition appearing again and again across different conditions.

This is what gives it stability. Not proof. Not fixed explanation. Continuity.

You begin to trust this. Not immediately.

At first, there is hesitation. A questioning of whether what you are sensing is reliable. Whether it can be depended on.

This is natural.

The mind seeks confirmation. It looks for something external to validate what is being felt.

Over time, something shifts. The need for confirmation softens.

Not because you have stopped questioning. Because the consistency of recognition becomes clear.

You have seen it enough times. Felt it enough times.

Recognised it across enough variation. That it no longer depends on a single moment of validation.

This is where knowing becomes quieter. Less reactive. Less urgent.

You are no longer trying to prove what you perceive. You are allowing it to remain as it is.

Open. Responsive. Continuing.

This does not mean everything is known. It does not remove uncertainty. It does not create a fixed understanding. It changes your relationship to what is unknown.

You are more able to remain with it. To allow it to unfold. To recognise that not everything needs to be resolved immediately.

This is what the bay has offered. Not answers.

A different way of knowing.

One that does not separate you from what is being known. One that does not require distance in order to understand. One that forms through participation. Through presence. Through attention that is sustained over time.

You are not outside of what you know. You are within it. This is what gives it its depth.

Not because it is complete. Because it continues.

Each time you return, it shifts slightly. Expands. Reorganises.

Not replacing what was known before. Carrying it forward.

This is how knowledge moves here. Not fixed. Not final.

Always in relation to what is present. Always influenced by what has come before. Always continuing.

You do not arrive at a final understanding. You move within it. You recognise more. You feel more. You respond differently.

This is not something you finish. It is something you live within. A different kind of knowing. Not held in the mind alone.

Held in the body. In perception. In the way you remain within what continues. And in that, something settles.

You no longer need to define everything in order to know it. You no longer need to hold it in place. You allow it to move.

And in allowing it to move, you remain connected to it. Not as something separate. As something you are already part of.

This is the knowing that forms here.

Quiet. Steady. Unresolved.

And unmistakably present.

Reflections - On Memory, Water, and Interpretation

This work has been formed at the edge of the bay.

Through return. Through attention. Through the body learning to recognise what continues.

What has been described here is not a theory to be proven. It is a way of perceiving.

A way of understanding how movement carries forward across water, across land, across life.

And it is important to be clear about what is being said. And what is not.

This work does not claim that water stores memory as fixed information in the way it is sometimes described.

It does not suggest that the ocean holds events as records that can be retrieved. It does not rely on ideas that cannot be supported.

What is being described is different. Water carries influence. It carries temperature, salinity, movement, and interaction forward through time.

Ocean currents extend across vast distances, connecting regions that appear separate. Conditions persist, not as stored data, but as ongoing movement.

In this way, memory is expressed through pattern, through continuation, through response. Not through storage.

This is supported by what is known about physical systems. Movement carries forward. Conditions influence what follows.

Nothing resets entirely from one moment to the next. The same is true of living systems.

Migration pathways continue across generations. Not as fixed instructions held in a single place, but as patterns maintained through interaction between body and environment.

Whales move through these pathways. Not retrieving a stored map. Responding to a field of conditions that carries continuity across time.

This is not abstract. It is observable. It is studied. It is ongoing.

The same principle can be recognised within the human body. Experience shapes response. The nervous system adapts. Perception becomes more refined through repeated exposure.

This is supported by research in neuroscience and epigenetics. The body does not store experience as a fixed narrative alone. It carries patterns of response. It becomes more able to recognise what it has encountered before.

This is what has been referred to here as cellular memory. Not memory as stored story. Memory as readiness.

Memory as response shaped by what has been experienced.

This is where the work of this book sits. At the meeting point between what can be measured and what can be felt.

Between scientific understanding and lived experience.

These are not opposing positions. They are different ways of engaging with the same continuity.

Science describes the mechanisms. Currents, migration, ecological systems, embodied cognition.

It provides structure. Clarity. Language.

This work describes the experience of being within those same conditions. Of sensing continuity directly. Of recognising pattern without needing to isolate it into a single variable.

Both are necessary. One does not replace the other.

This is not an argument for belief. It is an invitation to perception.

To recognise that memory can be understood not only as something held in place, but as something that moves.

Through water. Through Earth. Through life. And through the body.

This is why the language used throughout this book has been careful. To describe what is observed. To remain within what can be supported. To avoid overstating what cannot be known.

There are limits to what can be explained fully. Limits to what can be measured directly. This does not diminish what is experienced. It simply places it in context.

Recognition without proof is not the same as making unsupported claims. It is acknowledging that some forms of knowing arise through repeated experience.

Through consistency over time. Through the body learning to recognise what continues.

This kind of knowing is not fixed. It remains open. It adjusts as understanding deepens. It does not close the question. It keeps it alive.

This is where interpretation becomes important. What is felt must be held carefully. Not immediately turned into conclusion. Not extended beyond what can be supported.

The relationship between perception and explanation must remain clear. This is part of the discipline. To recognise without overstating. To feel without needing to define everything completely. To remain in inquiry.

The bay has offered a way to experience this directly. Not as an abstract idea. As a lived relationship.

Through water that moves. Through patterns that continue. Through the body learning to recognise what is already in motion.

This is not something separate from the wider understanding of the natural world. It is part of it. Aligned with it. Another way of engaging with it.

What remains is not a final conclusion. Not a statement that resolves everything.

Only this - Memory moves. It is carried through pattern, through movement, through response.

Across water. Across land. Across life.

And through the body that is capable of recognising it.

This is what has been offered here. Not as something to believe. As something to notice.

To return to. To remain with.

And to continue exploring, without needing to close what is still unfolding.

APPENDICES

Appendix A - The Practice of Listening Lines

This appendix offers a way to enter the work described throughout this book.

Not as a method to master. Not as something to complete. But as a quiet practice of returning.

Listening Lines are not something you find. They are something that form over time - through attention, through presence, through the body.

They are not visible in the way we are used to seeing. But they become known through repetition. Through noticing. Through relationship.

What follows is a simple way to begin.

1. Choose a Place

A Listening Line begins with place.
Not a place chosen for its beauty.
Not a place that feels extraordinary.
But a place you can return to.

It may be:

- A stretch of shoreline
- A set of rocks
- A path along water
- A quiet edge of land
- Even a place within your home that receives light, wind, or sound in a consistent way

The place itself is not the point.

The return is.

2. Arrive Without Purpose

When you come to your place, let there be no task.
No goal.
No outcome to achieve.
No need to understand anything.
Sit.
Stand.
Enter the water if that is part of your practice.
But arrive without trying to make something happen.
This is not observation in the analytical sense.
It is not data collection.
It is a form of being-with.

3. Stay Longer Than Feels Necessary

Listening Lines form through time.
Not dramatic time.
Not extended retreats. But through staying just a little longer than you normally would.

Long enough for the surface layer of attention to soften.
Long enough for the need to “do something” to pass.
Often, it is after this point that something shifts.

Not in the place - but in you.

4. Let the Body Lead

This is not a thinking practice.
It is a bodily one.

Notice:

- Temperature
- Movement
- Tension
- Breath
- Subtle shifts in posture
- The way your body responds to what is around you

The body recognises patterns long before the mind names them.

This is where Listening Lines begin.

5. Return Again

A single visit does not form a Listening Line.
Return is what creates it.

Come back:

- At different times of day
- In different weather
- Across seasons

You may begin to notice:

- Patterns of movement
- Rhythms of change
- Familiar responses within yourself

Over time, what once felt like "the same place" begins to reveal variation.

And what once felt unfamiliar begins to feel known.

6. Notice What Persists

Listening Lines are not only about what changes.
They are also about what remains.

Across time, begin to notice:

- What continues
- What returns
- What feels recognisable, even when conditions shift

This is not memory in the way we usually think of it.
It is not stored information.
It is something closer to continuity.
A pattern that holds, even as everything moves.

7. Resist the Urge to Conclude

There is no final insight required here.
No moment where the practice is complete.
No single meaning to extract.
Listening Lines are not about arriving at an answer.
They are about remaining in relationship.

8. Let Devotion Form Quietly

Over time, something may begin to shift.
The place may feel different - not because it has changed, but because your way of being with it has.

You may find:

- You move more slowly
- You notice more, without trying
- You feel less separate from what is around you

This is not something to force.

It is what happens when attention becomes steady.
This is what devotion looks like here.

Not dramatic. Not declared.

But lived.

----

do not need to do this perfectly.

You do not need to do it often.

Even a small act of return begins the line.

And once begun, it continues - through the body, through attention, through the quiet recognition that forms over time.

Appendix B - Memory Currents (Concept Overview)

This appendix offers a way to understand the idea that sits beneath this work.

Not as a fixed definition. Not as a theory to prove.

But as a way of describing something that may already be felt through practice.

In Appendix A, the reader is invited into a simple act of return.

To sit. To notice. To remain.

Over time, something begins to shift. Not in the place itself. But in how it is known.

This is where the idea of *memory currents* begins.

What Is Meant by "Memory"

The word memory is often used to describe storage. Something held. Something retained. Something that can be retrieved.

This is not how the word is being used here. In this work, memory is not treated as something fixed or contained within a single object or location.

It is understood as something that moves. Not stored. But expressed.

Not held in place. But recognised through pattern.

Memory as Pattern and Continuity

Across natural systems, patterns repeat.
Tides move in cycles.

Ocean currents follow established pathways.
Marine species migrate along consistent routes across vast distances.

These patterns are observable, measurable, and well documented.

Research from organisations such as CSIRO and marine ecology groups including the Dolphin Research Institute continues to map and study:

- Ocean currents
- Coastal ecosystems
- Species behaviour and migration
- Environmental change over time

What becomes clear through this body of work is not that memory is stored in a literal sense - but that **patterns persist**.

This persistence - across time, condition, and change - is what this work refers to as *memory currents*.

The Role of the Body in Recognition

Scientific observation measures and records. The body recognises.

Through repeated return, the body begins to register:

- Subtle differences

- Familiar conditions
- Rhythms of change
- Patterns of movement

These recognitions often occur before language.
Before naming.
Before interpretation.
This is not memory as recall.
It is memory as *recognition*.
A knowing that forms through relationship and repetition.

Observation and Interpretation

There is an important distinction within this work.

What is observed:

- Movement of water
- Presence or absence of species
- Changes in light, wind, and temperature
- Seasonal and tidal variation

What is interpreted:

- Meaning assigned to those patterns
- Connections drawn across time
- Questions about continuity, behaviour, and response

This work moves between both. It observes where possible. And it reflects where meaning begins to form. These are not presented as the same thing.

On Water and Memory

There are ongoing discussions - scientific, cultural, and philosophical - about the idea of water and memory. This work does not attempt to prove or disprove those claims.

Instead, it remains grounded in what can be experienced and observed:

- Repetition
- Pattern
- Continuity
- Response

The term *memory currents* is used as a way of describing these qualities.

Not as a scientific assertion. But as a language of relationship.

Continuity Across Deep Time

Along the Victorian coastline, fossil evidence reveals that whale species have inhabited these waters across vast geological timescales.

Research published by Museums Victoria, including findings from Beaumaris, identifies ancient whale species such as the Pygmy right whale. These findings do not suggest continuity in the sense of individual memory.

But they do demonstrate:

- Long-standing presence
- Evolution across time

- Repeated occupation of coastal environments

In the present day, whale species continue to move along the Australian coastline.

The connection between past and present is not framed here as retained memory. But as continuity. Life moving through time.

Patterns emerging, disappearing, and reappearing in new forms.

Holding the Question

At times, events occur that do not fit easily within known patterns. Whales entering enclosed bays. Unexpected movements. Moments that draw attention.

These events can be recorded. They can be studied. But they are not always fully explained.

This work does not attempt to resolve these moments. Instead, it allows space for a question to remain ... Whether such events are incidental or part of a continuity not yet fully understood is not concluded here.

Appendix C - Port Phillip Bay: Environmental Context

This appendix provides a grounded overview of the place that sits at the centre of this work.

Not to explain it fully. Not to define it completely.

But to offer context for what is observed, experienced, and returned to throughout these pages.

Before the Bay

Port Phillip Bay has not always existed in its current form. Geological evidence indicates that the area now known as Port Phillip Bay was once a land basin - a low-lying plain shaped by ancient river systems, marshlands, and shifting coastlines.

Over time, rising sea levels - particularly following the last ice age - flooded this basin, forming the bay as it is known today.

What now appears as a contained body of water is, in geological terms, relatively recent.
And yet, it carries layers of deep time beneath it.

A Living Marine System

Today, Port Phillip Bay is a dynamic and diverse marine environment.

It includes:

- Seagrass meadows
- Rocky reefs

- Sandy seabeds
- Intertidal zones
- Coastal wetlands

These environments support a wide range of marine life.

Research and monitoring by organisations such as CSIRO and local marine groups including the Dolphin Research Institute continue to document:

- Biodiversity within the bay
- Changes in water quality
- Species behaviour and habitat use
- Long-term environmental shifts

This is not a static environment.

It is one of ongoing interaction - between land, water, climate, and life.

Sanctuaries and Coastal Life

Within the bay are protected areas that offer insight into its ecological richness.

At Jawbone Marine Sanctuary, what begins as a simple shoreline of sand and rock reveals, just beneath the surface, a complex and active marine habitat.

Seagrasses, reef structures, and marine species coexist in ways that are not always visible from above.

Similarly, at Ricketts Point Marine Sanctuary, shallow reef systems support a diversity of life that shifts with tide, light, and season.

These places demonstrate something important - what appears quiet or ordinary at the surface often holds far more activity beneath.

Species Within the Bay

The bay is home to a wide range of species.

Among them are:

- Fish and invertebrates that inhabit reef and seagrass environments
- Birdlife along the shoreline and wetlands
- Marine mammals, including dolphins

Pods of Common bottlenose dolphin are known to inhabit and move through the bay. Their presence is studied and monitored, yet still retains elements that are not fully predictable.

At the northern end of the bay, the colony of Little penguins at St Kilda Breakwater continues to grow.

This colony exists on a human-made structure. It is an example of something often overlooked - that human presence does not always remove life from a place. At times, it becomes part of the conditions within which life continues.

Movement, Exchange, and the Bay Entrance

Port Phillip Bay is not isolated.

It connects to the Southern Ocean through a narrow entrance known as The Rip.

This opening governs:

- Water exchange
- Tidal flow
- Movement of nutrients and species

The entrance is dynamic and at times unpredictable.
It allows for movement in and out of the bay -
but not without constraint.

This balance between openness and containment shapes much of the bay's character.

Human Interaction and Living Systems

The shoreline of Port Phillip Bay is deeply shaped by human presence. Piers, breakwaters, ports, and coastal developments form part of the environment.

And yet, over time, these structures can become integrated into the ecosystem.

Marine growth forms on their surfaces. Species adapt to their presence. Habitats emerge where none previously existed.

This does not remove responsibility. If anything, it deepens it.

To recognise that human-made environments are not separate from nature but are part of the conditions that life responds to.

What Is Seen - and What Is Not

Much of what defines the bay is not immediately visible.

From the shoreline, it can appear still.

Even uneventful.

But beneath the surface:

- Currents move
- Species interact
- Systems shift continuously

Observation depends on how one looks.
And how often one returns.

Port Phillip Bay is not a fixed environment.

It is a living system shaped by:

- Geological history
- Ongoing ecological processes
- Movement between ocean and land
- Human interaction over time

To sit beside it, or within it, is to be in relationship with something that is continuously changing.

And yet, through repeated attention, patterns begin to emerge. Not imposed. But recognised.

This appendix does not attempt to capture the bay in full.

Only to offer enough context to understand that what is experienced through practice is grounded in a real, observable, and evolving place.

Appendix D - Whale Migration (Regional Overview)

This appendix outlines what is currently known about whale movement along the Australian coastline.
It does not attempt to interpret behaviour beyond what is observed.

It does not seek to explain what is not yet understood.
It offers a framework of movement - within which certain moments, including those described in this work, can be placed.

Migration Along the Australian Coast

Whale migration is one of the most extensively studied large-scale animal movements in the marine environment. Along the eastern coastline of Australia, populations of the Humpback whale travel annually between feeding and breeding grounds.

These journeys span thousands of kilometres.

Movement is generally understood as:

- South to north migration (May to August) toward warmer waters for breeding and calving
- North to south return (August to November) toward colder Antarctic feeding grounds

These routes are consistent across generations.
They are tracked through long-term observation, tagging studies, and ongoing research.

Southern Waters and Coastal Presence

In southern Australian waters, including areas such as the Great Australian Bight, the Southern right whale is known to gather in coastal regions.

These areas provide:

- Sheltered conditions
- Suitable environments for calving
- Proximity to shallow coastal waters

Unlike humpbacks, Southern Right whales often remain within southern regions rather than undertaking long northward migrations.

Their presence along southern coastlines is seasonal, yet relatively consistent.

Feeding and Breeding Cycles

Whale migration is closely linked to the distribution of food and the requirements of reproduction.

Broadly:

- Feeding occurs in colder, nutrient-rich waters where krill and other food sources are abundant
- Breeding and calving occur in warmer waters, where conditions are more suitable for newborn calves.

This separation of feeding and breeding grounds is a defining feature of whale migration patterns.
It reflects ecological necessity rather than variation in behaviour.

Movement Beyond the Expected

While migration pathways are well documented, variations do occur.

These may include:

- Changes in timing
- Deviations from established routes
- Occasional entry into environments not typically associated with migration pathways

Among these are rare instances of whales entering enclosed or semi-enclosed bays.
Including, at times, Port Phillip Bay.

Such events are:

- Observed
- Recorded
- Not common

They are not considered part of standard migratory behaviour.

And in many cases, they are not fully explained.

Continuity Across Time

Whale presence along the southern Australian coastline is not limited to contemporary observation.

Fossil evidence from locations such as Beaumaris indicates that whale species have existed in these regions across deep geological time.

Research from Museums Victoria identifies species including the Pygmy right whale within these fossil records.

These findings demonstrate:

- Long-standing marine presence
- Evolutionary continuity
- The persistence of large marine mammals within these environments across time

They do not suggest direct continuity of individual behaviour.

But they do establish that these coastlines have been part of whale existence for far longer than current observation alone reflects.

What Is Known - and What Remains Open

Current scientific understanding provides:

- Established migration routes
- Clear feeding and breeding patterns
- Increasingly detailed tracking of whale movement

At the same time, there remain aspects of whale behaviour that are not fully understood.

Particularly in relation to:

- Deviations from migration routes
- Rare entries into enclosed environments
- Individual or group variation in movement

These moments are documented.

But they are not always explained.

Holding the Question

This appendix does not seek to resolve these unknowns.
Only to acknowledge them.

Within a framework where much is known,
there are still moments that remain open.

Whether such events are incidental
or part of a continuity not yet fully understood
is not concluded here.

Closing Note

Whale migration demonstrates:

- Scale
- Consistency
- Endurance across time

It is one of the clearest examples of patterned movement within the natural world.

And yet, within that pattern, there is variation.

Moments that fall outside expectation.
Moments that draw attention.

This appendix provides the known structure of movement. The rest of this work sits alongside it - not in opposition, but in quiet observation of what sometimes occurs within it.

References

This work draws on a combination of lived observation, place-based practice, and established scientific and environmental research.

The following sources support the environmental, ecological, and historical context referenced throughout the book.

Scientific and Environmental Organisations

- CSIRO
  Marine and Atmospheric Research, Ocean Currents, Coastal Systems, and Climate Data
  https://www.csiro.au

- Dolphin Research Institute
  Port Phillip Bay Dolphin Populations, Marine Behaviour Studies, Conservation Programs
  https://www.dolphinresearch.org.au

- Parks Victoria
  Marine National Parks and Sanctuaries - including Port Phillip Heads Marine National Park
  https://www.park.vic.gov.au

- Environment Protection Authority Victoria
  Water quality monitoring and environmental health of Port Phillip Bay
  https://www.epa.vic.gov.au

- Melbourne Water
  Catchment systems, waterways, and coastal environmental data
  https://www.melbournewater.com.au

Geological and Fossil Research

- Museums Victoria
  Melbourne Fossil Find Reveals Deep Past of World's Most Mysterious Living Whale
  https://museumsvictoria.com.au/media-releases/melbourne-fossil-find-reveals-deep-past-of-world-s-most-mysterious-living-whale

- Beaumaris Bay Fossil Site (Victoria, Australia)
  Marine fossil records including ancient whale species and seabed formations
  https://www.theage.com.au/national/victoria/meet-the-man-who-s-found-so-many-fossils-in-melbourne-he-s-opened-his-own-museum-20260304-p5o7d6.html

Whale Migration and Marine Biology

- Australian Government Department of Climate Change, Energy, the Environment and Water
  Whale and dolphin conservation, migration data, marine protection
  https://www.dcceew.gov.au

- International Whaling Commission
  Global whale migration research and conservation frameworks
  https://iwc.int

- Australian Antarctic Division
  Southern Ocean ecosystems and whale feeding grounds
  https://www.antarctica.gov.au

Species References

- Humpback whale
  https://www.dcceew.gov.au/environment/marine/marine-species/whales-dolphins/humpback-whale

- Southern right whale
  https://www.dcceew.gov.au/environment/marine/marine-species/whales-dolphins/southern-right-whale

- Pygmy right whale
  https://australian.museum/learn/animals/mammals/pygmy-right-whale

- Common bottlenose dolphin
  https://www.dolphinresearch.org.au

- Little penguin
  https://www.park.vic.gov.au/places-to-see/sites/st-kilda-pier-and-breakwater

Port Phillip Bay - Key Locations

- Port Phillip Bay
  https://www.park.vic.gov.au/places-to-see/parks/port-phillip

- Jawbone Marine Sanctuary https://www.park.vic.gov.au/places-to-see/parks/jawbone-marine-sanctuary

- Ricketts Point Marine Sanctuary https://www.park.vic.gov.au/places-to-see/parks/ricketts-point-marine-sanctuary

- Pope's Eye https://www.park.vic.gov.au/places-to-see/parks/port-phillip-heads-marine-national-park

- Chinaman's Hat https://www.park.vic.gov.au/places-to-see/parks/port-phillip-heads-marine-national-park

- St Kilda Breakwater https://www.park.vic.gov.au/places-to-see/sites/st-kilda-pier-and-breakwater

- The Rip https://www.park.vic.gov.au/places-to-see/parks/port-phillip-heads-marine-national-park

Local Environmental Knowledge and Observational Context

- Long-term shoreline observation and immersion practices within Port Phillip Bay
- Repeated site-based engagement across shoreline, reef, and offshore environments
- Public environmental education materials and interpretive signage across marine sanctuary locations

- Community awareness and conservation efforts relating to marine life within the bay.

Additional Context

This work also draws upon:

- Established knowledge of tidal systems and ocean currents
- Marine ecology and coastal environmental studies
- Publicly available environmental and marine biology research
- Observational understanding developed through repeated return to place.

Note on Interpretation

Where scientific research is referenced, it has been used to support observable patterns and environmental context.

Where interpretation extends beyond current scientific explanation, it is presented clearly as reflection rather than fact.

This distinction has been maintained throughout the work to ensure clarity, integrity, and respect for both scientific knowledge and lived experience.

Acknowledgements

This work began with an unexpected moment.

A quiet disruption. A shift in attention.

A special acknowledgement is given to the two whales who entered the waters of Williamstown in 2025.

Your presence was not planned. Not explained. And not easily understood. But it was noticed. And it was felt.

Your arrival did not change the bay -
but it changed how it was known.
This work exists because of that moment.

To the shoreline of Williamstown, and the waters of Port Phillip Bay - thank you.
For holding, for shifting, for remaining.
For revealing what cannot be rushed.
For continuing, whether noticed or not.

To those who care for this place in visible and practical ways:

The teams and individuals within Hobsons Bay City Council, whose ongoing stewardship supports the health, access, and protection of the coastline.

The volunteers and community members of the Jawbone Marine Sanctuary Care Group, whose quiet and consistent work helps preserve what exists beneath the surface.

And the broader community of people who walk, swim, observe, protect, and return - often without recognition - yet contribute to the continuation of this environment through their presence and care.

To the researchers, scientists, and organisations whose work informs this book:

Those committed to understanding marine systems, species, and environments over time. Your work provides the grounding that allows observation and reflection to sit responsibly alongside what is known.

To those who have contributed to the broader understanding of these waters across time:

Through research, documentation, conservation, and education - both formally and informally. Your work forms part of what allows places like this to be understood, protected, and respected.

To the water itself - and all life within it:

Including the dolphins that move through these waters, the penguins that return to their shoreline homes, and the whales whose presence, at times, draws attention to something larger than any single moment.

And finally, to the act of return.

To the quiet, repeated choosing to sit, to notice, to remain.

This work exists because of that.

Final Thoughts

If you have read this far, then something in you has already begun to listen.

Not to the words - but to what sits beneath them.

This is not something you can finish by reading.
It only begins there.

If you want to understand what this work is pointing to, you will need to go to the water yourself.

Not once. Not briefly.

But enough times that something in you begins to recognise what remains.

The bay is not a place to visit. It is a place to return to. Again, and again - until what once felt separate begins to feel known.

This is how relationship forms. Not through explanation, but through continuity.

The bay does not require understanding. It asks for presence. It will not perform for you. It will not explain itself. But it will meet you - gently, and without force - if you are willing to stay.

The question is not whether the bay is there. It is whether you are.

If you are willing, begin your own Listening Lines.

Choose a place.

Return to it.

Sit.

Stand.

Enter the water, if that is available to you.

And do nothing except remain long enough for your attention to soften.

Let the place meet you in its own time. And allow what forms there to become your knowing.

Not as an idea.

But as relationship.

"I don't interpret the place.

I help you learn how to listen to it.

Now it is yours to begin."

Jen

ABOUT THE AUTHOR

Jenesis is a writer, spirit medicine woman, and guide devoted to helping you remember who you truly are and returning to living in harmony with the sacred rhythms of life. Through her books, ceremonies, comedy, art and teachings, she invites readers into deeper connection with themselves, the Earth, and the unseen wisdom that flows through all things.

She is the author of "She Is Sea: Water Lore, Ocean Medicine, and the Rising Sacred Feminine", "Whale Wisdom, Energy Body: Spiritual Intelligence and Chakra Alignment", "The Spiritual Imperative", "Listening Lines: Memory Currents – BELONGING" and more.

Her work explores the interplay between spiritual awakening and mastery, energy alignment, and the innate intelligence of nature, offering practical pathways back to sovereignty and wholeness.

Passionate about nurturing the next generation, Jen is also the author of the children's book "The Whale Who Forgot His Way: Swimming Back Home for His Birthday", which helps children explore intuition, self-trust, and the magic of their inner knowing.

Through storytelling, ritual, and sacred remembrance, she creates offerings that awaken the soul, reconnect us to the Earth, and inspire a more compassionate, conscious, and connected way of living.

Her motto:

"I help people reconnect with knowing that matters."

DISCOVER MORE ABOUT JEN'S OTHER BOOKS, ART, MUSIC, CEREMONIES, STAND UP COMEDY, SACRED RETREATS AND PUBLIC EVENTS BY VISITING

www.NewDreamingPublications.com

www.ingramcontent.com/pod-product-compliance
Lightning Source LLC
LaVergne TN
LVHW010659110826
845149LV00014B/3166